EXPLORING LIFE SCIENCES

EXPLORING LIFE SCIENCES

Archana Shukla

MD PUBLICATIONS PVT LTD
NEW DELHI
www.mdppl.com

Published by :

MD Publications Pvt Ltd
"MD House", 11, Darya Ganj,
New Delhi-110 002
Phone : +91-11-41563325 (Customer Service)
+91-11-41562846 (Editorial)
E-mail (orders) : order@mdppl.com
Website : www.mdppl.com

ISBN : 978-81-7533-190-7

Published and Printed by Mr. Pranav Gupta on behalf of **MD Publications Pvt Ltd** at Printext, New Delhi.

PREFACE

The life science is a challenging title as it examines the structure, function, growth, origin, evolution, distribution and classification of all living things. To understand science or to evaluate its impact on our lives, we shall have to know something about how it came to be as it is and what it has been in the past. And no one can give us this knowledge quite.

Ever since the Renaissance science has been growing exponentially, and today it is speeding along into the unknown faster than it has ever traveled before. New discoveries are altering our daily lives. Our outlook on the universe at large is being changed drastically, perhaps than we sometimes wish. But on the whole and with some reservations, we rather like science. We are certainly becoming more and more dependent on the products that science gives us. Materially, we are better off than any human generation has ever been, and the luxuries of earlier years have become our necessities. By our modern standards of convenience and comfort, the kings who reigned in the Middle Ages were underprivileged, and yet, in spite of its gifts, we are not entirely happy with our science. Our age truly the golden age of science has also been called the "Age of Anxiety." Even the scientists themselves have begun to question the human value of some of their activities. Some of the more thoughtful of them have begun to feel, perhaps a little vaguely, that they are missing out somewhere. Some are becoming a little troubled about the potential dangers of scientific discovery.

Now the dangers that accompany advances in science are just something we shall have to live with. Scientific discoveries cannot be stopped, because science has already achieved a certain degree of autonomy. No man or group of men can halt its course. Short of a world-wide catastrophe, we can imagine nothing that will bring scientific discoveries to an end, and we may look forward confidently both to new discoveries and to the new problems that they will raise. We know that we shall have new benefits and that we shall have to face new hazards.

Our growing scientific knowledge will give us more power because knowledge as a rule leads to power. But power in itself is always neutral. Expressed in the language of the past, power can be used for either good or evil. How we use our power will depend upon our wisdom, and human wisdom, we have to admit, is not unlimited. Nor has science reached a point where it can increase our wisdom or make it adequate for our needs. But, even so, science is not completely helpless. Science has given us and will continue to give us reliable data on which to base our reasoning and thus it may make our limited wisdom more effective. It can, if we do not obstruct it, expose the fallacies in our wishful, partisan, or simplistic thinking. And this brings us to one aspect of the history of science that deserves attention.

Often our scientific discoveries are not welcomed. Science not only uncovers new and unexpected truths, it also exposes ancient and often well-loved errors. As it advances, it forces us to learn but it also forces us to unlearn and this latter process is always painful. Almost no one likes to be forced to rearrange his intellectual equipment and this may account for the hostility that science still evokes. We have long recognized a persistent resentment toward science among the *pagani* and, during the past two centuries, among more educated groups. Even the scientists themselves are not completely immune to the unsettling effects of new discoveries, and the study of past scientific controversies has shown, beyond any reasonable doubt, that scientists are often intensely human.

But over and above the limitations that good scientists suffer as members of our species in good standing, they have been acquiring an added handicap of steadily increasing severity. For the past century, they have been forced into a growing specialization that continually narrows their fields of competence. True, the scientist of today knows far more than his predecessors have ever known, in comparison with the total to be known he knows much less. Percentage wise (a terrible expression) he is falling behind and his knowledge covers a more restricted range. The whole forest of the known tends to recede into a distant and undifferentiated background, against which the scientist now concentrates his attention on a sharp, clear and microscopic structure of some tiny twig.

I am very grateful to Mr. Pranav Gupta, Director of MD Publications Pvt Ltd, New Delhi for his cooperation and support for this book.

Archana Shukla

CONTENTS

1

Life Science

An Introduction

Definition and Meaning

Oparin's words seem strikingly current in light of the recent interest in the possibility of life on Mars. We know two things about life's origin. First, as Oparin pointed out in 1924, life originated in the absence of life and in the absence of free oxygen. Second, the appearance of life on Earth was apparently not a lengthy process. The earliest bacteria appeared almost as soon as Earth's crust was cool enough to support life. Oparin felt that "a billion years are needed to realize" life's origin from inorganic precursors, but the geological record does not allow this much time for what might have been the first event of spontaneous generation. If the recent claims of ancient Martian life are true, then either life planet-skipped by some sort of phenomenon of panspermia or life is very easy to create under the proper physical and chemical conditions.

In our solar system at least, life could not get a foothold on a planet until the megacratering crisis had ended. This crisis was the period of early bombardment of planets by planetesimals (gigantic, subplanetarysized meteors). An accretion of meteors such as these formed the planet in the first place. So much energy was released with each incoming rocky mass that the entire planetary surface was melted and presumably sterilized. This era of meltdowns has been called the "impact frustration of life" and ended on earth with the end of the intensive period of megacratering (as indicated by the ages of craters on the moon) about 3.8 billion years ago.

Rocks struck by meteoric impacts become pervasively fractured. When these fractures became fluid-filled, their surface areas expanded greatly, and thus may have become ideal sites, precisely the micro chemical factories needed for the origin of life. From a biological point of view, the fractures formed by incoming meteors represented a megacratering opportunity rather than a crisis.

The earliest life must have been microbial, the first forms probably being about .005 millimeter in diameter. A fascinating question concerning the origin of life is, "When did the first cell acquire the ability to distinguish self from nonself?" As A. G. Cairns-Smith argued in *Seven Clues to the Origin of Life,* life's origin may well have been as much a mineralogical phenomenon as a biochemical phenomenon. In his view, a crystalline form of life (Gene-1) gave rise to a fully "organic" form of life (Gene-2).

Cairns-Smith felt that there must have been some sort of inorganic scaffolding on which the earliest life would have started. He proposed clay as the living crystal of Gene-1. More recent research has shown that clay does not have the properties needed to act as the scaffolding of Gene-2. Nevertheless, the main biomolecular constituents of life (nucleic acids, proteins, and phospholipids) are the products of complex biochemical synthesis pathways that cannot have arisen, de novo, on their own. As with self-supporting stones in a stone archway, some sort of scaffolding must have supported the stones during construction.

The idea of earliest life lacking individualization, forming as something like a living crystal, an extended body form that permeated some special environment of Earth, is indeed attractive. But however the first cells came to be, life apparently remained unicellular for billions of years. Multicellular life, individuals composed of billions or trillions of cells, did not appear on the globe until long after life began.

The earliest organisms thought to represent multi cellular creatures are uninspiring as fossils, occurring as more or less shapeless organic films (carbonized impressions) on slabs of shale. The best that can be said about them, and this is by no means certain for all examples, is that they were eukaryotes, bearers of nucleated cells.

Eukaryotic cells are characterized by the presence of intracellular organelles, many of which were once free-living, and subsequently symbiotic, bacteria. This idea of a symbiotic origin for the organelles of eukaryotic cells gained momentum in the United States with Oparin's attendance at a conference in Wakulla Springs, Florida, in 1963. In the discussion session, Oparin presented the revolutionary idea of symbiogenesis, the thought that a new type of organism can emerge by the fusion of two unrelated types. This was the first time many of the Western conference participants had heard these ideas.

The American investigator Hans Ris, of Wisconsin, visited the Soviet Union and has advanced an idea similar to what was expounded several years ago in Russia by Mereshkovskii, namely, that a cell represents a symbiotic structure. They said that for the time being the idea was rather too audacious. But it is possible you could develop it in the direction of representing the formation of cells as a gradual association, aggregation of symbionts.

Hans Ris was Lynn Margulis's adviser in college; to my surprise, before I mentioned it to her in June 1996, she had never heard that he had visited Russia. There appears to be a fascinating and untold story about the development of symbiogenesis theory in Soviet Russia, a story that may have its share of Cold War intrigues. However, I am not surprised by Oparin's comments because he was one of the few scientists of his stature at the time to have had more than a passing familiarity with symbiogenesis theory. The only other was the great Russian geologist Vladimir Ivanovich Vernadsky (1863–1945), who studied under symbiogeneticist Andrei S. Famintsyn, "founder of the Russian school of plant physiology, who demonstrated the possibility of photosynthesis in artificial light." Vernadsky used his knowledge of symbiogenesis to found the now burgeoning field of biogeochemistry. In Vernadsky's view, biological processes are so important for our planet that it may truly be said that "life makes geology."

As Douglas R. Weiner points out in his review of Liya Nikolaevna Khakhina's book *Concepts of Symbiogenesis: A Historical and Critical Study of the Research of Russian Botanists* (translated into English in 1992), symbiogenesis is integral to the Russian traditions in the history of science. Andrei S. Famintsyn (descended from a sixteenth century Scottish immigrant whose name is the Russian translation of Thompson) sought to supplement Darwinism with symbiogenesis theory. Konstantin S. Mereshkovskii tried to displace Darwinism with his new symbiogenesis theory between 1900 and 1920. Boris M. KozoPolyanskii tried to incorporate symbiogenesis smoothly within the overall schema of Darwinian evolution. Khakhina explains the slow headway symbiogenesis theory made in most scientific circles outside Russia. She describes it in terms of the perception that through the 1950s, symbiogenesis did not accord with the prevailing explanations of evolution.

The idea of a symbiotic origin of organelles is now the accepted theory presented in biology courses throughout the world. Nevertheless, scientists who espouse symbiogenesis raise hackles among their colleagues in evolutionary biology. One response to the murmuring is to boldly point out that there are indeed problems with the 1950s explanation of evolution, commonly called the neo-darwinian modern synthesis. A strong case can be made that neo-darwinism is due for an intellectual shakeup. As we will see, the solutions to the mysteries of Ediacara will play an important role in updating the modern synthesis. We start at the beginning of the Ediacaran fossil record.

The first large, complex, unquestionably multicellular fossils appear about 600 million years ago in stratified rocks of northern Mexico. Complex life on land, recognized by my wife Dianna and me as the biogeophysical entity Hypersea, appears some 200 million years later.

Hypersea is the sum of eukaryotic life on land and all its symbionts. Despite its geological youth, Hypersea overwhelms the marine biota in terms of both total biomass and total biodiversity. This happens because the fluid connections between eukaryotes on land (particularly the ones involving plants and their root or mycorrhizal fungi) lead to a pumping of nutrients from the soil up into the photosynthetic parts of plants, a phenomenon we call hypermarine upwelling. Oparin neatly anticipated our Hypersea theory, even hinting at hypermarine upwelling back in 1963: "Imagine that land life did not exist. From the standpoint of a jellyfish, life on dry land is sheer nonsense. Through a complex process of adaptation, of water exchange of circulation [sic], such a form of life was able to arise."

The first complex multicellulars and Hypersea are separated by the great divide in the geological time chart, the Precambrian-Cambrian boundary. This boundary is marked by what has been called the Cambrian breakthrough, the abrupt appearance of virtually all major types of skeleton-bearing animals. A robust and continuing evolutionary debate regarding this breakthrough involves two main questions. First, did all the skeletalized animals appear suddenly at this time (the bang hypothesis), or do they have long histories that happened to leave virtually no fossil record (the whimper hypothesis)? Some authors advocate the whimper, others the bang. The whimperers are forced to admit that there is a major evolutionary radiation at the beginning of the Cambrian, although they try to keep the perceived number of new phyla appearing at this time to a minimum. The bangers see the phyla developing rapidly, and some postulate an unusual genetic reorganization that happens only at this time and is frozen into place (the green genes hypothesis, a version of the bang hypothesis). The main problem with this putative fixing of particular gene expressions is that it is difficult or impossible to test scientifically.

The main proponent of the green genes hypothesis, James W. Valentine of the University of California at Berkeley, does not support the more extreme statements of his idea, and says that the elaboration of early animal genes "may have been necessary, but was not sufficient, to drive the evolutionary creativity of the Cambrian." His 25-year quest to explain the Cambrian explosion in terms of gene regulation has not yet met with unequivocal success. Each successive Valentine paper on this subject seems to say, "Here is the latest breakthrough in modern genetic research; it must have something to do with the Cambrian explosion!" However, I believe that the origin of the major gene complexes in animals, an interesting subject in itself, has no necessary connection to the Cambrian event, and in fact may have been completely decoupled from it, the major steps in the formation of the animal genetic code having been taken well before the Cambrian. There will be a better harvest for scientists among fossils and the ecological issues of the Garden of Ediacara.

Bang or whimper, the Cambrian armored animals include many of familiar types that can be placed in still extant phyla. But for at least 50 million years before the Cambrian explosion, there existed a marine world of large and unusual creatures.

These organisms constitute the Ediacaran biota. They have also been called the Ediacaran fauna, but because the term *fauna* implies animals, and paleontologists are not confident that all of the Ediacaran forms were animals; prudence requires the less specific term *Ediacaran biota,* or simply *Ediacarans.*

Diverse communities of multicellular creatures appear with the first members of the Ediacaran biota. My recent find in Mexico of trace fossils associated with the oldest Ediacarans indicates that true animals were unquestionably part of the biota. Also present were the Ediacaran body fossil forms, less easily classified.

The Ediacaran biota seems at first glance to be another case of apparent spontaneous generation. Oparin's billion years are not evident here. My field research indicates that the Ediacarans sprang forth, fully formed, without a long record of evolution. This leads to the second question.

How could this happen? Furthermore, what kind of creatures are represented by the Ediacarans? Were they the first animals? They certainly seem to be associated with trace fossil evidence of the earliest animals, but in the view of German invertebrate paleontologist Adolf Seilacher, they are not animals at all. In 1983 Seilacher destabilized what had been the consensus viewpoint (that is, Ediacarans as early animals) by pointing out that they had a quilted body architecture totally unlike anything seen in animals. Following insights made by German paleobotanist Hans D. Pflug, Seilacher argued that Ediacaran forms were *sui generis,* representatives of a group of high taxonomic rank that went extinct at the beginning of the Cambrian.

A well-known science writer, following Seilacher's story, called the Ediacaran forms "aliens here on earth," meaning that they represented an alien body form no longer represented in the world. Later work has demonstrated that these forms survived well into the Cambrian. However, the newer research has not settled the question of what these forms were, or how they fed. Many mysteries remain. The solutions may well involve a fuller understanding of the phenomenon of symbiogenesis. The question of the origin of life is an enduring puzzle, but we are just as ignorant about the origin of complex life.

Precisely eighteen years ago, this journal began carrying a regular feature initially entitled "Law and the Life Sciences." The column's title seemed especially apt since the Center itself was then still known officially by its since-forgotten name, the Institute of Society, Ethics, and the Life Sciences. The column was devoted to the issues generated by medicine and its associated scientific disciplines and technologies as they surfaced in the courts, legislatures, and other lawmaking bodies.

Some years later, in a foreword to a collection of columns and other essays by George Annas (this column's author for its first fifteen years), Daniel Callahan remembered that, at the time, his main doubt about a regular feature was whether "law and ethics" would provide enough material to be interesting "month after month."(1) In context, this observation was meant to be amusing or even ironic, not simply because Professor Annas had managed to find such fascinating things to say but because it was by then self-evident that there was no scarcity of cases, statutes, regulations, and other materials of interest to physicians, scientists, ethicists, and other readers of the Hastings Center Report.

Yet it is worth remembering that this state of affairs is not inevitable nor was it always the case. Indeed, in 1973 when I first compiled a book of materials called "Law and the Life Sciences" for use in law school classes at Columbia, Pennsylvania, and Yale, most of the contents "were either from other disciplines (such as sociology and philosophy) or were cases and statutes that did not deal directly with the life sciences but from which I could draw analogies by examining the possible application of rules in contracts, criminal law, constitutional law, or administrative law."(2) Today, whole books are devoted to subsets of the same territory and are filled with cases and statutes directly on point, with no need to borrow; the wildest hypotheticals can now be replaced with actual cases and statutes.

Law in Bioethics

What accounts for this change and for the persistence of troubling issues at the intersection of law, medicine, and ethics in the pages not just of the Report but of other specialized and general interest legal and medical journals?

One reason the journals and law books now bulge with bioethics cases and statutes is that the life sciences are more powerful than ever. Science has improved on Robert Sinsheimer's famous observation that we must look not to the stars but to our genes for the herald of our fate because scientists now have ways not just to know our genes but to engineer them. And the same is true across the spectrum of human life in the contemporary world. When death could not be prevented or even retarded much less, a concepts created there was little for ethicists or lawyers to worry about. Now that physicians and scientists have technologies to control (at minimum, manage) human life from conception to death, human choice replaces fate.

Such developments in the life sciences, as well as in the way health services are organized and financed, challenge existing social and professional norms; where those challenges are substantial and intractable, the people involved not infrequently turn to courts, legislatures, or executive agencies for help.

Furthermore, the questions at issue are so central to human life and to the character of our society—involving, as they do, individual liberty and privacy, fair treatment, family relations, personal identity, health, and life itself—it is hardly surprising that they not only lead to contentious cases but also become lightning rods for special interest groups. In the eyes of the latter, the manner in which the law treats biomedical issues (such as fetal research, abortion, "Baby Doe," contract motherhood, genetic engineering, euthanasia, animal experimentation, and health care access and rationing) can fundamentally define our society.

Yet the explosion of new biomedical capabilities and a heightened awareness of their importance need not necessarily mean a corresponding increase in lawmaking. In a society in which ethical standards were clear, extensive, and strong, there would be no need for the law at all. Thus, might it be that we have turned to the law to tell people what they must do because ethics has failed to convince them of what they should do?

A more modest appraisal would admit that ethics can be influential but should not always be expected to be dispositive. First, problems arise in bioethics precisely because new biomedical discoveries and capacities challenge existing definitions of humanness. For example, the use of mechanical means to sustain respiratory functions in what would otherwise be a corpse brings new urgency to the age-old question, What does it mean to be a human being? Moreover, new technologies (such as embryo transfer with contract pregnancy) can raise old issues (such as Who is a mother?) in such a novel fashion that the issue cannot be readily resolved by reference to the existing canon. It would not be surprising for ethics to be less dispositive in such circumstances, as people debate the meaning of ancient ethical principles for the new circumstances and perhaps even argue over which set of ethical principles or rules are applicable.

Second, even when a particular ethical system is detailed enough to provide a definitive answer to all the dilemmas posed by the life sciences, the system itself will not enjoy universal acceptance in a society as pluralistic as ours. Over the past two decades, the nation has not only received an unprecedented flood of immigrants from non-Western cultures but has also abandoned the aspirational myth of being a "melting pot" in which cultural differences dissolve into a generic American ethos. With the acceptance—and even celebration of multiculturalism has come a recognition that different value systems can produce very distinctive and diverse attitudes toward health and illness and the role of biomedical interventions.

Thus, for several reasons one need not adopt the stronger view that the law has played a dominant role because ethics lacks the ability to get people to do what they should do. Instead, one could conclude that our society turned to the law rather than relying on ethics alone to guide the life sciences because in the particular circumstances ethics did not provide a complete or widely enough accepted account of right conduct to command adherence of the relevant parties. Lacking a single ethos or set of unifying, inherited myths, "lawyers and law are what hold [America] together," as Joseph Campbell has observed.

Some Consequences of Law's Role

Whatever the explanation, there can be little doubt that the major role played by the law has had enormous consequences in the development of bioethics. Indeed, Professor Annas has gone so far as to suggest that "American law, not philosophy or medicine, is primarily responsible for the agenda, development, and current state of American bioethics."(4) Though I would not assign such dominance to any one factor, the law has without doubt strongly influenced the central focus of bioethics, the values of bioethics, and the methodology of bioethics. And—to the considerable extent that bioethics is an American invention and export—the influence of American law has been felt even in societies in which legal institutions play a less pronounced role than they do in the United States.

Much is made—often self-critically—of our reputation as a litigious society. The biomedical arena provides a good reminder why this inclination may, however, be a strength rather than a weakness. Power over how and by whom the decisions to use or forgo particular biomedical technologies will be made involves the sort of issues about life, liberty, and well-being that are the particular province of the law. We ought to be pleased to see such decisions made (or at least subjected to review) through due process of law rather than through the naked exercise of power and influence.

But as appropriate as it may be, turning to the courts and secondarily to legislatures and administrative agencies to curb usurption of authority by the powerful has consequences for the way the issues are framed. Take, for example, abortion, a central topic in bioethics. Had religious and philosophical ethics been the major shaper of public discourse on this subject, the central issue obviously would not be as it now is whether physicians and their pregnant patients have a right to make decisions about abortion without being unduly burdened by state restrictions.

Furthermore, framing the issue this way has not only altered the development of the law (placing more authority in the courts and foreclosing the sort of legislative developments that have occurred in most other Western nations) but has also stunted the discussion of the fundamental ethical issues raised by the practice. In particular, the pro-choice forces are reticent about discussing circumstances when abortion would not be the correct choice, lest in doing so they seem to waver in their commitment to the hard-won recognition of a constitutional right to abort.

The common law's focus on rights is heightened in the American legal system by our Constitution, which enshrines rights not only as one of the counterweights against centralized power but ultimately to protect individuals against the tyranny of democratic majorities at all levels of government. Even powerful private actors—of which health care professions and institutions are prominent embodiments today—are subject to duties and restraints when their actions imperil certain personal interests recognized by common law and the Constitution. Although valid objections can be raised to casting all bioethics dilemmas in terms of competing rights, "rights are crucial to deal with power inequality even when our preference would be to conceive of relationships in terms of caring and connection. This tension has been a recurring theme in law and bioethics."

Plainly, the concepts employed by courts in recognizing such rights of patients as informed consent and the privacy of medical information were enriched by philosophers' elaborations of personal autonomy, just as the significant limitations and privileges that hedge in these rights drew on such ethical principles as beneficence and nonmaleficence. But the law often led the way, both in its individualistic focus on rights (in preference, for example, to notions of virture or community or the mutual ties inherent in relationships) and in fleshing out the content of those rights.

Explicit and Implicit Values

A ready example of such content is the law's attachment to self-determination. When applied in medical cases, this position has the obvious consequence of leaving injured patients to bear the bad as well as the good results if the risks of accepting or declining interventions were properly disclosed and they agreed to proceed. But this orientation also has the less obvious consequence of taking off the table questions about the nature of the choices themselves. That is, focusing on freedom to choose among the alternatives made available allows us to ignore the question of whether other choices should be made available. Similarly, health law—in contrast to health ethics—has been more concerned with negative than with positive rights: people are to be protected from unwanted things being done to them by the health care system ("freedom from...") rather than being enabled to gain those things they want from the health care system ("freedom to...").

The American legal system's devotion to certain central values, such as equality, nondiscrimination, justice, and autonomy, has explicitly influenced the way bioethics has been developed and articulated. But the legal system's major role has also shaped the development of the field in implicit, subtle ways as well. Perhaps the clearest example of this lies in bioethics' assimilation of the law's love of process, sometimes to the exclusion of careful analysis of substance.

In his first column, George Annas criticized the New Jersey Supreme Court's then recent decision regarding Karen Quinlan as "one of the most wide-ranging—and potentially dangerous opinions in medical jurisprudence." Although the heart of Professor Annas's complaint was that the court unduly protected physicians from liability, he also took aim at the procedure that the court suggested hospitals should follow in reaching a decision to discontinue life support, namely having an "ethics committee" decide whether "there is no reasonable possibility of [the patient's] ever emerging from [his or her] present comatose condition to a cognitive, sapient, state." Rather than rejoice in the tendency of a committee to diffuse moral responsibility, the court should have been worried that it becomes too easy to decide to allow death when no one is in charge.

Another subtle influence that the law has had on bioethics arises from the adversarial posture of court cases, which poses all issues as dichotomies in zero-sum settings. Again, this not only denies the possibility of mutual advantage but also removes from the calculus the interests of others who are not parties.

Effects on Bioethics Methodology

Besides shaping the focus and values of bioethics, the law has also influenced the means through which bioethics is developed and articulated. Although the field does not have a single methodology, its connection to a series of renowned court cases—from the Nuremberg trials of the Nazi concentration camp physicians to the disputes over forgoing life-sustaining treatment for Quinlan, Saikewicz, Conroy, Brophy, Perlmutter, and Cruzan, to the battles about Baby Fae, Baby Doe, and Baby M—has linked bioethics closely to the common law's reliance on fact specific, incremental reasoning.

While bioethics has often been described in terms of midlevel ethical principles (such as autonomy, beneficence, and justice), the establishment of those principles—and, more importantly, the articulation of the way they should be related, combined, and overridden when they conflict—has occurred through an inductive process that owes its inspiration at least in part to the way judges resolve the signal bioethics cases that come before them. This common-law methodology is similar to that used in casuistry, "the analysis and classification of 'cases' and 'circumstances,'" which Albert Jonsen and Stephen Toulmin have urged is the appropriate method for bioethics. The legal method has the advantage over casuistry in resting on a generally accepted authority (the courts) that legitimates the designation of particular paradigmatic examples as applicable to the resolution of new cases, whereas it has been suggested that bioethics itself cannot be assumed to arise from a common morality or to have an agreed-upon moral authority.

Besides being case-oriented and incremental, the method that the law has contributed to bioethics also takes into account that we meet as "moral strangers" who do not "share enough of a concrete morality to allow the common discovery of the basis for the correct resolution of a moral controversy." The law's focus on individuals allows bioethics to negotiate around this moral separateness and pluralism by deferring to each individual to the extent possible. The central danger with this posture, however, is that bioethics could become obsessed with giving individuals due process because that is easier, as George Annas has pointed out, than the frustrating but essential task of developing substantive rules.

Whether or not bioethics succeeds in rising to this need, one can be sure that it will continue to be fed by a strong and ever-growing stream of cases and statutes. The greatest challenge for the law and legal commentators as regards bioethics in the next eighteen years may be to help the field tackle the broader social issues of access to and organization of care; since these issues involve community well-being and not solely individual rights, they have generally fallen outside the case-orientation of traditional medical law.

How much direct effect difference of climate, food, etc., produces on any being is extremely doubtful. My impression is, that the effect is extremely small in the case of animals, but perhaps rather more in that of plants. We may, at least, safely conclude that such influences cannot have produced the many striking and complex co-adaptations of structure between one organic being and another, which we see everywhere throughout nature. Some little influence may be attributed to climate, food, etc.: thus, E. Forbes speaks confidently that shells at their southern limit, and when living in shallow water, are more brightly coloured than those of the same species further north or from greater depths. Gould believes that birds of the same species are more brightly coloured under a clear atmosphere, than when living on islands or near the coast. So with insects, Wollaston is convinced that residence near the sea affects their colours. Moquin-Tandon gives a list of plants which when growing near the sea-shore have their leaves in some degree fleshy, though not elsewhere fleshy. Several other such cases could be given.

The fact of varieties of one species, when they range into the zone of habitation of other species, often acquiring in a very slight degree some of the characters of such species, accords with our view that species of all kinds are only well-marked and permanent varieties. Thus the species of shells which are confined to tropical and shallow seas are generally brighter coloured than those confined to cold and deeper seas. The birds which are confined to continents are, according to Mr Gould, brighter-coloured than those of islands. The insect species confined to sea-coasts, as every collector knows, are often brassy or lurid. Plants which live exclusively on the sea-side are very apt to have fleshy leaves. He who believes in the creation of each species, will have to say that this shell, for instance, was created with bright colours for a warm sea; but that this other shell became bright-coloured by variation when it ranged into warmer or shallower waters.

When a variation is of the slightest use to a being, we cannot tell how much of it to attribute to the accumulative action of natural selection, and how much to the conditions of life. Thus, it is well known to furriers that animals of the same species have thicker and better fur the more severe the climate is under which they have lived; but who can tell how much of this difference may be due to the warmest-clad individuals having been favoured and preserved during many generations, and how much to the direct action of the severe climate? for it would appear that climate has some direct action on the hair of our domestic quadrupeds.

Instances could be given of the same variety being produced under conditions of life as different as can well be conceived; and, on the other hand, of different varieties being produced from the same species under the same conditions. Such facts show how indirectly the conditions of life act. Again, innumerable instances are known to every naturalist of species keeping true, or not varying at all, although living under the most opposite climates. Such considerations as these incline me to lay very little weight on the direct action of the conditions of life. Indirectly, as already remarked, they seem to play an important part in affecting the reproductive system, and in thus inducing variability; and natural selection will then accumulate all profitable variations, however slight, until they become plainly developed and appreciable by us.

Effects of Use and Disuse – There can be little doubt that use in our domestic animals strengthens and enlarges certain parts, and disuse diminishes them; and that such modifications are inherited. Under free nature, we can have no standard of comparison, by which to judge of the effects of long-continued use or disuse, for we know not the parent-forms; but many animals have structures which can be explained by the effects of disuse. As Professor Owen has remarked, there is no greater anomaly in nature than a bird that cannot fly; yet there are several in this state. The logger-headed duck of South America can only flap along the surface of the water, and has its wings in nearly the same condition as the domestic Aylesbury duck. As the larger ground-feeding birds seldom take flight except to escape danger, I believe that the nearly wingless condition of several birds, which now inhabit or have lately inhabited several oceanic islands, tenanted by no beast of prey, has been caused by disuse. The ostrich indeed inhabits continents and is exposed to danger from which it cannot escape by flight, but by kicking it can defend itself from enemies, as well as any of the smaller quadrupeds. We may imagine that the early progenitor of the ostrich had habits like those of a bustard, and that as natural selection increased in successive generations the size and weight of its body, its legs were used more, and its wings less, until they became incapable of flight.

Kirby has remarked (and I have observed the same fact) that the anterior tarsi, or feet, of many male dung-feeding beetles are very often broken off; he examined seventeen specimens in his own collection, and not one had even a relic left. In the Onites apelles the tarsi are so habitually lost, that the insect has been described as not having them. In some other genera they are present, but in a rudimentary condition. In the Ateuchus or sacred beetle of the Egyptians, they are totally deficient. There is not sufficient evidence to induce me to believe that mutilations are ever inherited; and I should prefer explaining the entire absence of the anterior tarsi in Ateuchus, and their rudimentary condition in some other genera, by the long continued effects of disuse in their progenitors; for as the tarsi are almost always lost in many dung-feeding beetles, they must be lost early in life, and therefore cannot be much used by these insects.

In some cases we might easily put down to disuse modifications of structure which are wholly, or mainly, due to natural selection. Mr Wollaston has discovered the remarkable fact that 200 beetles, out of the 550 species inhabiting Madeira, are so far deficient in wings that they cannot fly; and that of the twenty-nine endemic genera, no less than twenty-three genera have all their species in this condition! Several facts, namely, that beetles in many parts of the world are frequently blown to sea and perish; that the beetles in Madeira, as observed by Mr Wollaston, lie much concealed, until the wind lulls and the sun shines; that the proportion of wingless beetles is larger on the exposed *Desertas* than in Madeira itself; and especially the extraordinary fact, so strongly insisted on by Mr Wollaston, of the almost entire absence of certain large groups of beetles, elsewhere excessively numerous, and which groups have habits of life almost necessitating frequent flight;—these several considerations have made me believe that the wingless condition of so many Madeira beetles is mainly due to the action of natural selection, but combined probably with disuse. For during thousands of successive generations each individual beetle which flew least, either from its wings having been ever so little less perfectly developed or from indolent habit, will have had the best chance of surviving from not being blown out to sea; and, on the other hand, those beetles which most readily took to flight would oftenest have been blown to sea and thus have been destroyed.

The insects in Madeira which are not ground-feeders, and which, as the flower-feeding *coleoptera* and *lepidoptera*, must habitually use their wings to gain their subsistence, have, as Mr Wollaston suspects, their wings not at all reduced, but even enlarged. This is quite compatible with the action of natural selection. For when a new insect first arrived on the island, the tendency of natural selection to enlarge or to reduce the wings, would depend on whether a greater number of individuals were saved by successfully battling with the winds, or by giving up the attempt and rarely or never flying. As with mariners shipwrecked near a coast, it would have been better for the good swimmers if they had been able to swim still further, whereas it would have been better for the bad swimmers if they had not been able to swim at all and had stuck to the wreck.

The eyes of moles and of some burrowing rodents are rudimentary in size, and in some cases are quite covered up by skin and fur. This state of the eyes is probably due to gradual reduction from disuse, but aided perhaps by natural selection. In South America, a burrowing rodent, the tuco-tuco, or Ctenomys, is even more subterranean in its habits than the mole; and I was assured by a Spaniard, who had often caught them, that they were frequently blind; one which I kept alive was certainly in this condition, the cause, as appeared on dissection, having been inflammation of the nictitating membrane. As frequent inflammation of the eyes must be injurious to any animal, and as eyes are certainly not indispensable to animals with subterranean habits, a reduction in their size with the adhesion of the eye-lids and growth of fur over them, might in such case be an advantage; and if so, natural selection would constantly aid the effects of disuse.

It is well known that several animals, belonging to the most different classes, which inhabit the caves of Styria and of Kentucky, are blind. In some of the crabs the foot-stalk for the eye remains, though the eye is gone; the stand for the telescope is there, though the telescope with its glasses has been lost. As it is difficult to imagine that eyes, though useless, could be in any way injurious to animals living in darkness, I attribute their loss wholly to disuse. In one of the blind animals, namely, the cave-rat, the eyes are of immense size; and Professor Silliman thought that it regained, after living some days in the light, some slight power of vision. In the same manner as in Madeira the wings of some of the insects have been enlarged, and the wings of others have been reduced by natural selection aided by use and disuse, so in the case of the cave-rat natural selection seems to have struggled with the loss of light and to have increased the size of the eyes; whereas with all the other inhabitants of the caves, disuse by itself seems to have done its work.

It is difficult to imagine conditions of life more similar than deep limestone caverns under a nearly similar climate; so that on the common view of the blind animals having been separately created for the American and European caverns, close similarity in their organisation and affinities might have been expected; but, as Schiödte and others have remarked, this is not the case, and the cave-insects of the two continents are not more closely allied than might have been anticipated from the general resemblance of the other inhabitants of North America and Europe. On my view we must suppose that American animals, having ordinary powers of vision, slowly migrated by successive generations from the outer world into the deeper and deeper recesses of the Kentucky caves, as did

European animals into the caves of Europe. We have some evidence of this gradation of habit; for, as Schiödte remarks, 'animals not far remote from ordinary forms, prepare the transition from light to darkness. Next follow those that are constructed for twilight; and, last of all, those destined for total darkness.' By the time that an animal had reached, after numberless generations, the deepest recesses, disuse will on this view have more or less perfectly obliterated its eyes, and natural selection will often have effected other changes, such as an increase in the length of the antennae or palpi, as a compensation for blindness. Notwithstanding such modifications, we might expect still to see in the cave-animals of America, affinities to the other inhabitants of that continent, and in those of Europe, to the inhabitants of the European continent. And this is the case with some of the American cave-animals, as I hear from Professor Dana; and some of the European cave-insects are very closely allied to those of the surrounding country. It would be most difficult to give any rational explanation of the affinities of the blind cave-animals to the other inhabitants of the two continents on the ordinary view of their independent creation. That several of the inhabitants of the caves of the Old and New Worlds should be closely related, we might expect from the well-known relationship of most of their other productions. Far from feeling any surprise that some of the cave-animals should be very anomalous, as Agassiz has remarked in regard to the blind fish, the Amblyopsis, and as is the case with the blind Proteus with reference to the reptiles of Europe, I am only surprised that more wrecks of ancient life have not been preserved, owing to the less severe competition to which the inhabitants of these dark abodes will probably have been exposed.

Acclimatisation —Habit is hereditary with plants, as in the period of flowering, in the amount of rain requisite for seeds to germinate, in the time of sleep, etc., and this leads me to say a few words on acclimatisation. As it is extremely common for species of the same genus to inhabit very hot and very cold countries, and as I believe that all the species of the same genus have descended from a single parent, if this view be correct, acclimatisation must be readily effected during long-continued descent. It is notorious that each species is adapted to the climate of its own home: species from an arctic or even from a temperate region cannot endure a tropical climate, or conversely. So again, many succulent plants cannot endure a damp climate. But the degree of adaptation of species to the climates under which they live is often overrated. We may infer this from our frequent inability to predict whether or not an imported plant will endure our climate, and from the number of plants and animals brought from warmer countries which here enjoy good health. We have reason to believe that species in a state of nature are limited in their ranges by the competition of other organic beings quite as much as, or more than, by adaptation to particular climates. But whether or not the adaptation be generally very close, we have evidence, in the case of some few plants, of their becoming, to a certain extent, naturally habituated to different temperatures, or becoming acclimatised: thus the pines and rhododendrons, raised from seed collected by Dr Hooker from trees growing at different heights on the Himalaya, were found in this country to possess different constitutional powers of resisting cold. Mr Thwaites informs me that he has observed similar facts in Ceylon, and analogous observations have been made by Mr H. C. Watson on European species of plants brought from the Azores to England. In regard to animals, several authentic cases could be given of species within historical times having largely extended their range from warmer to cooler latitudes, and conversely; but we do not positively know that these animals were strictly adapted to their native climate, but in all ordinary cases we assume such to be the case; nor do we know that they have subsequently become acclimatised to their new homes.

We must not, however, push the foregoing argument too far, on account of the probable origin of some of our domestic animals from several wild stocks: the blood, for instance, of a tropical and arctic wolf or wild dog may perhaps be mingled in our domestic breeds. The rat and mouse cannot be considered as domestic animals, but they have been transported by man to many parts of the world, and now have a far wider range than any other rodent, living free under the cold climate of Faroe in the north and of the Falklands in the south, and on many islands in the torrid zones. Hence I am inclined to look at adaptation to any special climate as a quality readily grafted on an innate wide flexibility of constitution, which is common to most animals. On this view, the capacity of enduring the most different climates by man himself and by his domestic animals, and such facts as that former species of the elephant and rhinoceros were capable of enduring a glacial climate, whereas the living species are now all tropical or sub-tropical in their habits, ought not to be looked at as anomalies, but merely as examples of a very common flexibility of constitution, brought, under peculiar circumstances, into play.

How much of the acclimatisation of species to any peculiar climate is due to mere habit, and how much to the natural selection of varieties having different innate constitutions, and how much to both means combined, is a very obscure question. That habit or custom has some influence I must believe, both from analogy, and from the incessant advice given in agricultural works, even in the ancient Encyclopaedias of China, to be very cautious in transposing animals from one district to another; for it is not likely that man should have succeeded in selecting so many breeds and sub-breeds with constitutions specially fitted for their own districts: the result must, I think, be due to habit. On the other hand, I can see no reason to doubt that natural selection will continually tend to preserve those individuals which are born with constitutions best adapted to their native countries.

2

Development of the Historiography of Science

Introduction

History of science is a relatively recent speciality, yet it has nonetheless an ancestry stretching back several centuries. This history of the historiography of science has its own intrinsic fascination, but also possesses a wider value. By examining the sequence, growth and proliferation of historical writing on science, it is possible to achieve a sense of the lineage and formation of history of science as a field of enquiry and scholarship. In achieving this sense, it is also possible to discern the emergence of the typical kinds of understanding, communication and expression whose development has been crucial in shaping the practices which now collectively constitute the disciplinary existence of historians of science. In what follows, therefore, we will be pursuing in a preliminary fashion the story in whose course first appear many of the basic themes, topics and interpretations which form the subject-matter of this *Companion*.

Historical Starting-Points

When does the historiography of science start? There is a genuine problem in assigning definitive, ascertainable origins for the historiography of science. This is because any scientist, in his actual scientific work, already has an orientation towards the past. He will have been educated into certain intellectual traditions, certain scientific practices, and his work will inevitably be involved in extending, or perhaps breaking with these traditions and practices. These efforts can often involve overt consideration of selective aspects of his science's history, so that it is no exaggeration to say that history of science is itself present, and can often appear within, works of scientific research and teaching. Two examples, from the science of chemistry in the eighteenth century, can reinforce this point. When Antoine Lavoisier (1743–94) came to publish some of his earlier studies of gases, he supplemented the account of his own research with the recent history of researches in this area. Similarly, the influential Dutch physician and chemist Herman Boerhaave (1668–1738) would introduce his students to the study of chemistry in part through providing a brief history of the science, a teaching practice which was often followed by later chemists.

Intellectual disciplines of all kinds tend to generate and produce their own relatively informal histories in these ways, and they offer a convenient starting point, a locatable series of origins for the historiography of science. Yet if we merely remained with these starting-points, we would emerge with only a rather scattered and incomplete sequence of incipient histories of scientific disciplines and their research programmes. Some other perspective, of broader and deeper significance for the origins of historiography of science, can perhaps be found. This perspective should seek more than the partial and individual recognitions of history contained within the scientific teaching and research of such as Lavoisier and Boerhaave, important and revealing though such examples are. But where might more significant origins be found, and of what might they consist?

They consist firstly in the recognition that science is not just a sequence of separate disciplinary activities, each with a discrete historical existence, but is also and much more importantly an activity possessing a general significance with respect to the overall course of human history itself. This recognition of the global significance of science, its world-historical importance, is part of the spectrum of justification for science which was promoted during the Scientific Revolution of the seventeenth century, most notably at the hands of the philosopher-scientist Francis Bacon (1561–1626). According to

Bacon, it is the aim of science to discover 'the knowledge of causes, and secret motions of things; and the enlarging of the bounds of Human Empire, to the effecting of all things possible'. This was a succinct, confident and ambitious statement of the nature and purpose of science, and its focus brought together the notions of scientific knowledge, power and progress. For Bacon and his followers in the seventeenth century, the implications of such an attitude were directed towards the future history of the human race, for whom science offered a progressive future based upon the power over nature which science now promised. Bacon and the other scientific revolutionaries of his period were therefore less concerned with elaborating upon science's past than with promoting an image of science which disengaged human history from its past and brought it into a new epoch. Thus, to discern the global origins of the historiography of science, we need in addition to specify that moment when this Baconian conception of science becomes itself an object of interest and enquiry for the historical consciousness of the West, as it seeks to understand and explain this distinctive feature of its historical existence.

Developments during the Enlightenment

We can date the emergence of this specifically historical concern with the global and epochal significance of science to the middle decades of the eighteenth century. It is associated with the intellectual movement which dominated that time, known to historians as the Enlightenment. The Enlightenment was essentially a programme of reform produced by philosophers and scientists dedicated to changing the intellectual, political and social terrain upon which humans had hitherto been obliged to live. They sought in particular a greater degree of individual political liberty and social equality than currently existed. A key to this attainment was intellectual liberation, and it was this presupposition of the Enlightenment which rendered science central to its aspirations, for Enlightenment thinkers erected science as the model of what the unfettered human intellect could achieve. The work of Galileo, Descartes, Bacon and Newton was mobilised as exemplary, as producing authentic knowledge of nature. This authentic knowledge not only liberated the human mind from the shackles of superstitious religion and outmoded metaphysics, but could be turned to productive material uses which would enhance prosperity, and thereby guarantee political and social progress. To make this claim, which echoed many of the original Baconian claims, Enlightenment intellectuals needed to pay a kind of detailed attention to the history of science which it had hitherto not received, for the claim was not simply a repetition of the original Baconian promise, but rather that this promise, since the time of Bacon and Galileo, had been partially fulfilled, and that this fulfillment justified the Enlightenment's progressive optimism. It was the Enlightenment, therefore, which first constructed and launched upon the world a historically-based view of science's intellectual, political and social significance for humanity. In the course of so doing, it laid down a series of assumptions concerning science and its historical existence which have been so influential that *all* Western historians of science have been formed within them. This holds equally whether historians have been persuaded of the Enlightenment's commitments, or whether they have attempted to modify or overthrow them.

Scientific Revolution

These assumptions can be seen emerging with particular clarity in the 1750s and 1760s. It was then that certain fundamental narratives, subjects and structures were developed in forms which became foundational for the historiography of science. In narrative terms, the Enlightenment wrote the history of what we now call canonically the Scientific Revolution. This may be read, for example, in attenuated form, in Jean d'Alembert's *Preliminary Discourse to the Encyclopedia of Diderot* (1751), where d'Alembert (1717–83) rehearsed a story which led from Galileo's trial by the Inquisition, through Bacon, Descartes, Kepler and Huyghens to Newton and Locke, a sequence whereby 'a few great men prepared from afar the light which gradually, by imperceptible degrees, would illuminate the world'. It was a history of intellectual emancipation from the influence of politically and spiritually repressive forces. The protagonists in this narrative take on heroic and exemplary status for d'Alembert and his readers: 'Such are the principal geniuses that the human mind ought to regard as its masters'. Short as d'Alembert 's narrative is, it nonetheless embodies features which continue to appear in Western accounts of this historical phenomenon. Firstly, it sets it within a bounded historical period, the seventeenth century. Secondly, it takes the events under view as connected and developmental, forming a coherent narrative unit. Thirdly, it presents these events as a significantly progressive contrast with what had preceded them. Fourthly, the developments have a fundamentally revolutionary nature. Fifthly, they are the mental products of individual men of genius who form the cast of characters essential to the narrative. Sixthly, they include developments in philosophy as well as science. Finally, they are intellectually authoritative.

Many of these features may now seem unremarkable as an account of the Scientific Revolution. This is precisely a measure of the success of the Enlightenment's invention of the Scientific Revolution: the elements it selected and to which it gave a coherent narrative expression now appear so natural to our understanding of the origin of modern science that we find it very difficult to imagine any account functioning without those elements. It is as well to be reminded, therefore, that d'Alembert's narrative is a humanly selected and constructed interpretation undertaken for particular purposes, and in principle as alterable as any other such interpretation.

The *History of Astronomy* by Adam Smith (1723–90) was written at much the same time as d'Alembert's *Preliminary Discourse,* and illuminates another and rather different foundational tendency in Western historiography of science. This tendency is referred to in the full title of Smith's work: *The Principles which Lead and Direct Philosophical Enquiries; Illustrated by the History of Astronomy* (1795). Smith's work was by no means a straightforward disciplinary history of astronomy. Although it treated the development of astronomical thought from ancient times up until Isaac Newton, this development was subsidiary to his main interest, which was focused upon the universal principles by which the human mind understands and explains the natural world. Smith saw the human mind as constantly attempting to produce simple, unified and coherent representations of nature. It is provoked into this exercise by the perception of unusual, anomalous observations which do not fit with conventional expectations of how nature behaves. Faced with such observations, the mind produces a novel set of ideas, a scientific theory, which satisfactorily explains the disturbing observations. Although the detail of Smith's views bears some comparison with more modern philosophical accounts of science and scientific change, it is more significant in other ways. In Adam Smith we find a rather complex intertwining of history of science with philosophical concerns, in the course of which a view of the developmental or progressive view of the nature of science is expressed. Smith created a specific role for history of science in relation to the broader philosophical attempt to understand what general principles, if any, underlie and structure the course of scientific development.

In so doing, Smith was forging a link between the history and philosophy of science which has proved to be of enduring importance. Its continuing importance can be seen in works which have made a major impact on historiography of science in the nineteenth century, such as those of William Whewell (1794–1866), and in the twentieth century, such as those of Thomas Kuhn. Historiography of science, in other words, has from its origins onwards often accompanied a more specifically philosophical quest for understanding the nature of science, and this feature has in turn often affected the historiography, lending it a philosophical vocabulary and motivation. In this too, historiography of science is a child of the Enlightenment, which sought not only to make science the pre-eminent form of intellectual activity, but sought also within the human mind and within human history to discover the principles of scientific reasoning and progressive scientific development.

Contribution of Joseph Priestley

This preliminary survey of the Enlightenment's historiography of science would not be complete without attention to the historical work of Joseph Priestley (1733–1804), England's most famous man of science in the latter half of the eighteenth century. Priestley wrote, among other historical works, a work entitled *The History and Present State of Electricity* (1767). Less polemical than d'Alembert's, less philosophical than Smith's, Priestley's historiography is nonetheless just as revealing. Although its historical focus is confined largely to the development of electrical science in the eighteenth century, it avowedly wished to illustrate the theme illustrated in their different ways by d'Alembert and Smith, namely the progress of the human mind: 'It is here that we see the human understanding to its greatest advantage, increasing its own powers and directing them to the accomplishment of its own views; whereby the security and happiness of mankind are daily improved'. As such, historiography of science is preferable to, more instructive and delightful than histories of politics and warfare. Historiography is given here a directly educative, moralising role by Priestley, but perhaps more interesting is the way he conceives his subject matter, since it is here for the first time that historiography of science portrays science definitively as an active, highly instrumentalised, above all *experimental* activity. This contrasts notably with the versions of science portrayed by d'Alembert and Smith, where history of science is the history of mind, of intellectual ideas. Priestley, himself an instrument-maker and experimenter of note, understood science and its progress in less exalted terms. It was a history of practical discoveries made through experiment, aided by the construction of appropriate material apparatus. Discovery was innovative for Priestley, but constituted by small steps, a gradual series of improvements.

Negative Aspect of Scientific Historiography

Thus far, we can see how the Enlightenment set in place basic features of the historiography of the origins of modern science and installed within it liberal notions of freedom, progress and individualist creativity. Additionally, it set forth a structural connection between historiography of science and philosophy, and it pioneered a historiographical focus on innovative experimental discovery. These are all features of major import for the development of historiography of science, but it would be wrong to leave the Enlightenment without noticing one further feature which its historical understanding of science produced. Common to all aspects of Enlightenment historiography so far has been a deeply positive evaluation of science. Yet the Enlightenment also produced negative evaluations of science's role in history. The philosopher Jean Jacques Rousseau (1712–88), in his *Discourse on the Moral Effects of the Arts and Sciences* (1750) characterised science not as promoting freedom, but as substantially negating it.

So long as government and law provide for the security and well-being of men in their common life, the arts, literature, and the sciences, less despotic though perhaps more powerful, fling garlands of flowers over the chains which weigh them down. They stifle in men's breasts that sense of original liberty, for which they seem to have been born; cause them to love their own slavery, and so make of them what is called a civilized people.

Here, the civilised pursuit of knowledge and culture, with which Enlightened progress is identified, now becomes more sinister in its historical effects. It distracts men from the pursuit of true liberty, and at the same time both obscures the recognition of oppression while propping up the agents and instruments of oppression. The example of Rousseau could be added to. It makes the point that the Enlightenment's historiography of science did not finally produce a unified, single-valued representation of science's historical significance. Rather, it produced a split conception, with positive and negative aspects which seem incapable of being brought together. And in this too, our own contemporary historiography is an authentic descendant of the Enlightenment.

Scientific Societies and Disciplinary Histories

Between the 1780s and 1830s the principal historiographical development which occurred was arguably related to the forms of development which science itself was undergoing at this time. Hitherto science had existed institutionally in universities and in learned societies such as the Royal Academy of Sciences in Paris and the Royal Society of London. The later eighteenth and early nineteenth centuries witnessed the growth of disciplinary-based scientific societies, whose members were devoted to the pursuit of one particular scientific discipline, such as geology or astronomy, rather than to the pursuit of natural philosophy in general. Science itself was therefore undergoing a process of division of labour which produced increasingly specialised disciplinary orientations. The historiography of science itself responded to these developmental features of scientific life, producing a number of disciplinary histories. These were noteworthy not simply for their recognition of coherent disciplinary formations, and for their attempts to create discrete and unified histories of scientific disciplines, complete with founding figures, fundamental innovations and so forth. As we have already seen, there had by this time long existed a tendency for particular sciences to generate their own informal histories in particular circumstances of research and teaching. More noteworthy was the way in which the construction of a disciplinary history could itself make visible and begin to define basic problems of historical understanding which in one form or another still pertain, and still continue to preoccupy disciplinary historians. When and how does a specialised discipline come coherently together? Does the process mark a cumulative development of pre-existing elements, or else mark a definite break with what went before?

Historiography of Chemistry

The most enduring location of such questions lay, and still lies, in the historiography of chemistry, for historians of chemistry are compelled to consider the problem of the science's historical lineage with respect to the preexisting practice of alchemy. This represents a key problem, for alchemy can be seen as an embarrassingly 'pre-scientific' activity, a jargon-ridden, secretive practice with impossible aims. Yet alchemists were also considerable experimenters, and developed theories of chemical 'elements', both of which features mean that, in historical terms, chemistry cannot simply be cut off from its alchemical past. This historical problem, of the emergence of the authentic scientific discipline in relation to its problematic past, was tackled for example in Thomas Thomson's *History of Chemistry* (1830–1). It also recurs within historical writing on the recently emerged science of geology, where the attempts of Charles Lyell (1797–1875) to re-establish the methodological and conceptual foundations of geological science entailed a historical designation and account of its origin. Such historical reflection on the 'genuine' origins often tends to accompany the emergence of particular scientific specialities. It is this process which constitutes the formulation of a discipline's identity, along with specification of subject-matter, methods, techniques and theories. While disciplinary historiography of science has, since the early nineteenth century, achieved a professional distance from such direct identity formulations, the nature and form of the problem of originary identity must always continue to haunt its efforts.

'Historicism'

For intellectual historians generally, the nineteenth century is often seen as the period in which historiography as a whole took on its modern form, producing many classics of historical writing. This itself can be seen as an outgrowth of the nineteenth century's devotion to what is termed 'historicism'. 'Historicism' is that view of man, nature and society which insists that all these are formed by processes of development through time, rather than by abstract, eternal static principles such as reason or justice. The two pre-eminent exemplars of nineteenth-century historicism are Gottfried Hegel (1770–1831) and Karl Marx (1818–83), both united in seeing human existence as being fundamentally produced by historical change. They adopted, however, radically different views of what constituted the dynamic core of historical development. For Hegel, development was essentially mental in nature, the growth of the human mind and human society towards full and rational self-consciousness, a position known as 'idealism'. For Marx, by contrast, the basic form of historical development was material, economic production. As men produce the essential material features of their life, they also derivatively produce their social and political relations, their consciousness, learning and culture, a position known as 'historical materialism'.

Work of William Whewell

These two forms of historiographical approach have been profoundly influential, for historiography generally and for certain aspects of the historiography of science in particular, and these aspects will be examined in more detail below. For the time being, however, our question is the impact of historicism upon the nineteenth-century historiography of science. Did it produce a home-grown equivalent of Hegel or Marx? The most likely pretender to such a title was the Cambridge-based scientist, philosopher and historian, William Whewell. Whewell produced a massive and erudite *History of the Inductive Sciences* (1837). It incorporated two features we have designated as already foundational for historiography of science. Firstly, it was a history of *sciences,* which divided up the historical world of science into the development of discrete scientific disciplines, such as astronomy, physical science, geology and so forth. Indeed, Whewell pushed this policy of division and subdivision to hitherto unheard-of lengths, inventing neologisms such as 'geological dynamics', 'thermotics' (the science of heat) and 'atmology' (the study of vapour). This process of division and subdivision created a representation of scientific development as an endless series of continual branchings, proliferating as history advanced towards the present day. Secondly, Whewell 's history was also specifically tied to an explicitly philosophical project, which indeed appeared as *The Philosophy of the Inductive Sciences* (1840). The history was to function as the essential background to and basis of an analysis of the principles of progressive scientific reasoning. As a mobilisation of history in the service of philosophy, it has never been surpassed in its thoroughness, which is itself an index of Whewell 's historicist leanings.

Whewell 's overall developmental picture of history of science is very much an extension of the Enlightenment's. It saw science's origins in the speculative works of Greek philosophy, defective in their lack of factual content; the Middle Ages were a 'stationary period', lacking progressive elements because weighed down by intellectual dogmatism and mystical forms of knowledge such as astrology and alchemy. Genuine scientific development was then resumed in the sixteenth and seventeenth centuries with Copernican astronomy and Newtonian physics. From that time successive scientific specialities had emerged and grown to maturity.

Whewell, however, considerably added to this basic picture, and not only with respect to the sheer amount of detail his work contained. Present in Whewell's work was a far greater methodological sophistication and narrative complexity than ever before. Methodologically, he insisted that for science to develop authentically, pre-conditions were required: in this case, the coexistent presence of both facts and theories, each being necessary for the progressive existence of the other. He also exemplified what has since become a standard technical method in studies which conjoin history with philosophy of science, the method known as 'rational reconstruction'. This method, rather than following a strictly factual chronological narrative of the emergence of a discovery or theory, attempts instead to reconstruct the process of rational developmental relations which, it is held, pertain to the discovery or theory's emergence. Whewell's pioneering use of this technique can be seen in his account of Newton's discovery of the principle of universal gravitation. Whewell analysed this down into five logically and conceptually distinct constituent propositions, and his history portrayed the separate emergence of each proposition, the order and connection of emergence being disconnected from any thorough and coherent chronological analysis.

As well as these methodological innovations, Whewell also introduced higher levels of narrative complexity than any of his predecessors. He saw history of science developing in 'epochs', highly charged periods of progress identified usually with the work of one individual, such as Newton. But this individual's work was not simply the individual's sole creative mind confronting and explaining nature unaided. Rather, the individual existed in already historically formed circumstances, created by the relevant discoveries of preceding scientists, which constituted the 'prelude' to the 'epoch'. The epochal significance of an individual such as Newton then consisted in drawing a whole series of preexisting discoveries within one unified, generalised framework, re-arranging a diverse and perhaps disconnected series of discoveries as functions of a single scientific proposition or law. This account of fundamental scientific change represented it as non-revolutionary. Underlying apparently revolutionary changes was a cumulative sequence of changes, and their culmination smoothes out any apparent contradictions. Nothing of scientific value is lost as science moves forward. In so far as they are true, former discoveries are preserved and incorporated within the culminating, unifying development.

Nothing which was done was useless or unessential, though it ceases to be conspicuous and primary. Thus the final form of each science contains the substance of each of its preceding modifications; and all that was at any antecedent period discovered and established, ministers to the ultimate development of its proper branch of knowledge. Such previous doctrines may require to be made precise and definite, to have their superfluous and arbitrary portions expunged, to be expressed in new language, to be taken into the body of science by various processes; but they do not on such an account cease to be true doctrines, or to form a portion of our knowledge.

Whewell's historical exposition of this narrative principle preserves the notion of progress essential to science while discarding the idea of progress as being discontinuous or revolutionary in nature. It is a conservationist, consolatory image of change also, for nothing of significance is ever lost. But above all it is historicist in outlook, for each present moment of scientific development incorporates the truth and value of its past, and so is definitively formed and produced by its history.

Scientific Biography

The nineteenth, historicist, century was also a great age of biographical writing, and historiography of science was also responsive to this larger tendency of nineteenth-century culture. Biography was an extremely popular form of writing, and served to communicate to a wider than scientific public aspects of the historiography of science. David Brewster (1781–1868), the Scottish physicist, wrote popular biographies of Galileo, Tycho Brahe and Kepler, and also produced a monumental biography of Isaac Newton, a work only recently superseded, so considerable was its range and detail. Brewster's biographies of scientists still have historiographical interest, for they often focus upon a problem which retains technical and general significance for historians of science. This problem is most clearly delineated in Brewster's treatment of Newton's alchemical interests. These constituted problematic matter for Brewster, revealing as they did Newton's involvement in the intellectually contemptible and morally reprehensible practice of alchemy. Brewster himself was finally at a loss to produce any coherent explanation for Newton's alchemy, and the problem he faced, of understanding how Newton, the paragon of scientific rationality, also took seriously the mystical practices of alchemy, has never disappeared for Newtonian scholarship. Although approaches to the problem are now more sophisticated than anything Brewster could manage, the issue of how one integrates these apparently opposing aspects of Newton's work still persists.

Biography has become conventional for historians of science since the mid nineteenth century, continuing to provide a useful focus of research and writing. Its tendency is to stress the individualist element in historiography of science, namely, to see the unique agent of scientific development as the individual mind as it grapples with scientific problems. Biography however adds also a humanising element often lacking in histories of particular theories or sciences. Because biography takes the human life as its narrative unit, and because any successful biography needs to clarify the personal meanings which the protagonist's life and career held for him, it must focus on what a scientist's work means for him personally, how it expresses the scientist's personality in psychological and social terms. While the individualist aspect of biography can lead to over-emphasis on individual genius at the expense of more thoroughly historical explanation, and while biography after Freud can sometimes give way to reductive and implausible psychoanalytic explanations, it remains the case that with appropriate and sensitive handling, a biographical focus upon a scientist's work can provide an important range of insights which other kinds of historiography cannot. This, combined with the greater popularity of biography with a more general reading public, ensures the continuing survival of scientific biography.

Historiography of Science in the Twentieth Century

As historiography of science moved into the twentieth century, it moved into a radically new situation. Before the twentieth century, historiography of science was produced largely by scientists themselves, and philosophers. During the twentieth century, historiography of science has become an increasingly professionalised discipline, produced by people who practise history of science as a specialised academic occupation within universities and colleges. This process, it is important to realise at the outset, has not yet been total. It remains true that historians of science are not simply recruited from a straight career ladder consisting of an undergraduate degree in history of science, followed by an academic career of teaching and research. Historians of science often have a prior educational background in some other subject, a science, philosophy, sociology or history, before switching into history of science. Its practitioners often include workers in museums, and the membership of its associations includes individuals whose professional occupations are largely unrelated to the business of teaching and researching history of science. This means that history of science has a peculiarly open professional structure, which partially may account for its singularly responsive and diverse intellectual history in the twentieth century.

With these qualifications entered, history of science has nonetheless generated the kinds of academic forms, publications and networks, and professional associations, which conventionally characterise academic professions. It has learned societies, many specialised journals, and has university and college departments devoted to it, although the departments tend to be fewer in number and smaller in size than those of most other academic disciplines.

The growth of these departments, journals and associations occurred most notably after the Second World War, but it was a process already well in train during the first half of the century. The Collège de France in Paris had a short lived professorship devoted to history of science from 1892 to 1913, and although the professorship was suppressed, Paris remained an important centre for history of science from the late nineteenth century onwards, through the work of the chemist Marcelin Berthelot (1827–1907), the historian philosopher Paul Tannery (1843–1904), and later in the 1920s, 30s and 40s, the historians Helene Metzger (1889–1944) and Alexandre Koyre (1892–1964), who published both scholarly monographs and articles in journals such as *Archeion* and *Scientia.*

In 1912 the Belgian historian George Sarton (1884–1956) initiated and carried through plans to found *Isis*, which became and still remains the leading journal in the history of science. The importance of such journals is easy to underestimate, but they did more than simply provide a venue for the publication of scholarly work. Journals also provide a network of professional communication, helping thereby to create and solidify a bonding sense of community for scholars who might otherwise have remained isolated individuals or small cadres. They thus have a considerable significance for a nascent but still struggling academic discipline, such as history of science in the inter-war period.

Sarton's career and the vision which impelled it are revealing. History of science for Sarton was the only arena of human activity which unequivocally demonstrated the progress of mankind. Science was the religion of this progressive, secular humanist, who could declare, in his *Introduction to the History of Science* (1927–48) that 'The history of science is the history of mankind's unity, of its sublime purpose, of its gradual redemption'. Despite the intensity of his commitments, and his success in maintaining *Isis* as a viable project, Sarton's career nonetheless never managed to solidify itself institutionally. Moving to Harvard after the First World War, and undertaking some undergraduate teaching there over the next three decades, he nonetheless failed to persuade Harvard to support a department for the history of science. There were signs too that the involvement of scientists under the German Reich in the 1930s and

1940s shook his earlier faith in the history of science as the record of mankind's progressive redemption. That said, Sarton 's attempts to provide bibliographic resources for history of science, his commitment to laying down basic and professional methodological criteria for his subject and of course *Isis* itself, all remain enduring contributions to his chosen field.

Of more intellectual influence than Sarton was another emigre from Europe, the Russian Alexandre Koyre. Koyre was a historian of science whose work on seventeenth-century science, notably the *Etudes Galileennes* (1939) came to provide an intellectual model for many younger historians. Whereas Sarton had pursued a broad introductory overview, Koyre's scholarship was characterised by close textual exegesis of important scientific texts, carefully tracking their conceptual structures, to clarify the fundamental intellectual ideas which underlay the advances made by a Galileo or a Newton. Koyre's historiography has therefore a strongly idealist cast. For him, science was a kind of pure thought, approximate to philosophy, and Koyre himself approached scientific texts as a philosopher whose own philosophical commitments derived from the paramount idealist thinkers in the Western philosophical tradition, Plato and Hegel.

Koyre worked in France before the Second World War, then in New York during its course. From 1945 to 1964 he divided his time between Paris and American universities such as Harvard, Yale and Princeton. It was in this American setting that his work became particularly influential. This influence attained its full significance not only because it profoundly affected the approaches of a number of scholars who were to publish significant work in the 1950s, 1960s and 1970s. It was additionally significant because this period coincided with the notable professional expansion of history of science in higher education in America and elsewhere, witnessing an increasing number of academic programmes and departments being devoted to the history of science. It was in this setting, the coming-together of an expanding profession and a distinct historiographical approach, that Koyre's work took on central importance for the discipline and enabled him to have a structural effect upon modern historiography of science. That effect can be measured out in the work of influential American researchers and teachers such as Thomas Kuhn, Charles Gillispie and Richard

Westfall, a generation of scholars who have all variously acknowledged Koyre's profound intellectual impact.

If professional American historiography bears the decisive imprint of Koyre 's idealism at its formative stage, this by no means completes the description of recent and contemporary historiographical development. Materialist approaches, looking back ultimately to the work of Karl Marx, have also made considerable headway in the twentieth century. Historical materialism as applied to historiography of science, sees science as being produced and determined by the social and economic relations in which science takes place. Thus, instead of seeing science as a purely intellectual activity, developing according to its own, internal conceptual dynamics, historical materialism interprets science as a form of intellectual production, tied to the economic preoccupations, class interests and ideological values of particular historical periods and cultures.

A very basic form of this kind of work may be read, for example, in the Soviet historian Boris Hessen's account of Newton's *Principia,* which systematically relates the scientific content of Newton's work to specifically economic aspects of the society in which Newton lived. Since Hessen's work, materialist historiography of science has developed considerably more sophistication, and has resulted in major scholarly endeavours. Joseph Needham's huge *Science and Civilization in China* (1954–84) not only attempts to see science as something shaped by the culture which produces it, but in so doing has opened up to scholarly understanding a whole vast area with which western scholars have been unfamiliar. Histography of science tends very much to adopt a Eurocentric perspective, and Needham's work is a valuable corrective to the historical vision of things scientific which most Western scholars promote.

Professionalisation of History of Science

From historical materialism, and from other intellectual resources such as sociology, has developed a kind of historiography known as 'social history of science', which has had increasing impact within the discipline. Aspects of social history of science, and the resources which it draws upon, may be pursued in more detail in several of the articles which immediately follow in this section of the

Companion. Overall, social history of science incorporates several different kinds of study. It may rest content with providing a detailed history of science's institutional development at particular times and places, showing how particular communities of scientists are formed. It may also relate institutional development to wider social and political features of a culture or nation. From there, it may point out ways in which particular fields of science have been formed by, or become responsive to, features of social and political change. It may further point to the ways in which the content of certain scientific theories is produced by and contains ideological commitments typical of the society in which the theory is produced. It may point out how science itself affects the wider social, economic and political spheres. Much theoretical and practical labour has been devoted to such topics by historians, sociologists and philosophers of science over the last three decades. This work has not been uncontroversial. Although few would disagree that historical knowledge of science's social development is both useful and desirable, as is knowledge of science's impact upon society, social history of science can nonetheless question some of the cherished historical images produced originally by the Enlightenment and developed thereafter. Is science, after all, historically typified by the term 'progress'? The role of scientifically-produced technologies for military purposes and for ecologically-damaging manufacturing processes questions the suitability of terms such as 'progress' for understanding science's history. If science can be seen as responsive to social, economic, political and cultural forces and values, in what sense can it still be maintained that science produces authentic, neutral, objective knowledge? If scientists work typically in collective, institutionalised settings, should we continue to conceive of the individual scientist as the unique agent of scientific development? In short, social history of science tends overall to call in doubt the liberal certainties of progress, authenticity and individualism centralised by Enlightenment historiography and its nineteen hand twentieth-century successors. It may be, then, that historiography of science, in recently questioning its old, foundational assumptions, has initiated a revolution in its own historical understanding.

The increasing professionalisation of history of science has also aided in the abandonment of old historiographical assumptions. Two in particular deserve mention; firstly, the view that the Middle

Ages were devoid of scientific interest, and secondly the view that science, historically speaking, was an activity typified by its secular nature, which was held to contrast with the religious belief systems which preceded the rise of modern science. The scientist-philosopher Pierre Duhem (1861–1916) was among the first to question the view that the medieval centuries lacked any significant scientific development, showing in particular how medieval physical thought was subject to historical changes in ways which anticipated the sorts of changes usually attributed to the period of Galileo. The rich and variegated nature of medieval science has since been further revealed by scholars such as Crombie and Claggett. Historians have also had to come to terms with the impossibility of characterising science's history as something separate from religious belief and theological principle. Once there existed sufficient and detailed attention to the works and manuscripts of scientists such as Newton, Kepler, Boyle and many others in the seventeenth, eighteenth and nineteenth centuries, it became increasingly obvious that their scientific work could not simply be separated off from their religious beliefs and theological reasoning. Any sympathetic consideration of their science has to realise how it was integrated within the patterns of religious culture, rather than posing itself as an alternative to religion.

As well as these structural alterations in historiographical interpretation of the overall historical course and nature of Western science, the professionalisation of history of science has also and quite simply meant that far more of science's history has been discovered and written about in the last forty years than in the two preceding centuries. This process is incapable of summary, so diverse is the range of work that it covers, but in general terms what it has produced are increasing degrees of specialisation. Whereas for a Whewell or a Sarton it was still possible to write general histories of science, historians of science now tend to identify themselves in much more specialised terms: as historians of biology, or of chemistry, or of the social sciences; of American science or German science; of medieval science, or of early modern science, or of twentieth-century science. Among other things, the proliferation of specialised journals marks this process. This kind of specialisation allows much closer and more detailed scrutiny of scientific development than was ever possible for preprofessional or early professional practitioners.

Is historiography of science now therefore simply comparable to other disciplines, divided into narrowly focused specialities, and losing sight of larger issues and perspectives? This is significantly not the case, and for two reasons. Firstly, specialisation applies particularly to research, rather than to teaching. Because historians of science teach in rather small departments which are often obliged to offer general, introductory courses, they are equally obliged, as lecturers and tutors, to retain a general and updated awareness of the field as a whole, and to be professionally familiar with subjects as far apart as medieval optics and the Manhattan Project which produced the atomic bomb. This feature of professional life keeps in place an attention to the larger issues of scientific development.

Secondly, because history of science is an open professional structure which involves and includes practitioners of other disciplines such as philosophy, sociology and history, it is peculiarly sensitive and responsive to the kinds of information, theoretical perspectives and practical techniques which those other disciplines produce. Philosophers persist in producing new and different unifying theories of scientific development. Sociologists originate a variegated array of social approaches to science. Historians pioneer novel techniques of historical investigation. Many of these work their way, more or less rapidly, into the developing body of knowledge, technique and interpretation which make up the historiography of science. The field is now therefore constituted in a way which both allows fruitful specialisation and encourages generalist perspectives and a process of theoretical and methodological innovation, maintaining thereby its commitment to tackling the fundamental and global meanings of science in the history of the Western world.

3

History of Science and the Philosophy of Science

Introduction

Philosophers of science generally trace their roots to Aristotle's *Posterior Analytics*. That work with its emphasis on the structure of explanation, the nature of causation and the linkage between concepts and experience effectively paved the way for much of the subsequent development of the discipline. History of science is a comparative newcomer, usually seen as emerging from the pioneering studies of Priestley, Smith and Montucla in the late eighteenth century. Despite their quite divergent origins, each of these disciplines first came into its own during the nineteenth century. It was then, for instance, that philosophy of science began to assume an identity distinct from epistemology and when history of science came to be seen as more than an ancillary branch of the general history of learning.

During the great flowering of these disciplines in the century from the 1830s until the 1930s, the two subjects developed side-by-side, generally in very close liaison. Scholars such as William Whewell, John Herschel, Augustus De Morgan, Auguste Comte, Ernst Mach, Pierre Duhem, Abel Rey, Pierre Duhem, Emile Meyerson, E. A. Burtt, Arthur Hannequin, Kurt Lasswitz, Paul Tannery, Ernst

Cassirer and Alexandre Koyre happily intermingled historical and philosophical concerns in their research. They inclined to the view that science needed to be understood in cognitive terms, and that such an understanding must encompass both what science had been (hence the historical component) and what science ought to be (the philosophical element). Despite numerous disagreements about specific points of doctrine, all believed that there were important epistemological lessons to be learned from a detailed understanding of the genesis and development of scientific ideas.

Beginning sometime during the 1930s, however, each discipline began to go its own way. Philosophers of science, smitten with logical positivism's alluring promise of rigour, began to think that the method of conceptual analysis alone was sufficient to formulate an adequate understanding of the scientific enterprise; a detailed familiarity with the history of science (particularly of science before the twentieth century) was thought to have no particular bearing on questions about the conceptual and methodological foundations of the sciences. Significantly, none of the towering figures of logical empiricism (*e.g.*, Schlick, Carnap, Reichenbach and Hempel) saw any need to follow in the footsteps of Comte, Whewell, Mach and Duhem, who had insisted on the necessity of using history to illuminate and sometimes to adjudicate between rival philosophical doctrines. Formal, logical analysis tended to replace historical research as the preferred mode of presentation.

Historians of science, for their part, began to maintain that there was something perniciously 'Whiggish' about using historical materials to address general questions, philosophical or otherwise, about the nature of science. Historians of science from Whewell to Conant had formerly argued that the great virtue of the history of science is that it serves the cultural function of teaching us how science works and how scientific knowledge is wrought; but historians of science since the 1940s and 1950s have generally been chary about using the historical record to address broad or general questions about the nature of scientific change and about the justification of science. By the late 1950s, therefore, history and philosophy of science had become as remote from one another's concerns as they had once been intimately intertwined. Historians thought it inappropriate to address epistemological questions in their research; philosophers saw no need to 'stoop' to consulting the historical record.

There emerged a small but influential handful of scholars (*e.g.*, Feyerabend, Hanson, Buchdahl, Toulmin and Kuhn) who sought to persuade philosophers to re-think their dismissal of the relevance of historical research to philosophy of science. It had become clear by then that the positivist account of science was a gross misconstrual of science. Explanation in science did not typically satisfy the formal models of the positivists. Scientists did not draw the line between theory and observation in anything like the way that positivists had supposed. Empiricist theories of empirical support and confirmation apart from generating their own internal contradictions - had dismally failed to account for the manner in which scientists used evidence and made theory choices. For such reasons, it was suggested that the philosophy of science associated with logical positivism was ill-suited to provide any epistemic rationale for real science.

As a result of this critique, philosophical research in the last two decades has seen the return to respectability, if not yet to dominance, of the older tradition of pursuing philosophical and historical researches in tandem. Shapere, Laudan, McMullin and Lakatos, along with many others, are heirs to that tradition. The return of philosophers of science to historical concerns has been prompted by many factors, but the two chief among them are: (1) a belief that processes of theory, change and temporal progress are among the central epistemic determinants of science; and (2) a realisation that the *justification* of philosophical claims about how science works rests in part on the adequacy of those claims to actual science. One might have thought that the former factor would be especially potent as a basis for renewing the ties between historians and philosophers of science, since an account of scientific change must be, *inter alia*, a general theory about the historical development of science. And so it has been for philosophers, who now recognise that 'historically-orientated philosophy of science' is one of the major genres for approaching philosophical questions.

By contrast, and despite the emergence of theory change as a central problem on the philosophers' agenda, the attitudes of many historians of science towards the philosophy of science have hardened considerably, to the point where historians of science nowadays venture at their peril (and on pain of compromising their reputations) to address overt epistemological questions in their research. Prestigious journals in the history of science fail to carry articles which express explicit philosophical concerns, and many historians of science (especially in the English-speaking scholarly community) go to considerable lengths to disavow an interest in philosophical questions, and adamantly deny the mutual relevance of historical and philosophical research.

If my claims are correct about the hostility of historians to engaging the philosophical questions which have traditionally motivated the historical study of science, then the interesting case of Thomas Kuhn and the reception of his work by his fellow historians of science may be instructive. Here is a scholar who has doubtless had a greater impact on the broader intellectual community than any other historian of science in our century, perhaps even more than any general historian of his generation. Moreover, his central works – *The Structure of Scientific Revolutions* and *The Essential Tension*– are quite open about addressing themes of a general and philosophical sort about theory change. His handling of those themes has been widely cited by social and natural scientists, philosophers and literary theorists. Yet, reading the current scholarly literature of the history of science, one would scarcely be aware of the existence of Kuhn as a theorist of science. I know of no other active discipline which, over the course of almost a generation, has managed so systematically to fail to address the ideas of its most influential theorist. And what goes for Kuhn is just as valid for all the other theorists of scientific change, whether card-carrying historians of science (such as Holton or Cohen) or philosophical interlopers (like Lakatos, Toulmin, Feyerabend and a host of others).

As a result, relations between history and philosophy of science are sorely strained. Philosophers of science, at least many in their ranks, have become convinced that the history and philosophy of science make sense only if conducted in consort. By contrast, the prevailing view among historians of science is to the effect that the philosophical suitor proposing matrimony should be summarily dismissed.

All of this raises a key question, namely, whether there is a coherent rationale for historians to resist so tenaciously a marriage which the philosophers thought had already been consummated.

Interestingly, there is virtually no discussion in print, which explains why historians are resisting the proposal. One can point to no classic essay or monograph which shows definitively why the call of Kuhn, Crombie, Conant and Koyre (echoing the earlier call of Whewell, Mach and Duhem) for historians of science to engage issues in the theory of knowledge has fallen on such deaf ears. In the absence of written sources, one must depend on the oral tradition to find out why historians are unmoved by the clarion call. That means that a part of my analysis will be evidentially suspect, for, since historians have not grappled with this issue in print, I have had to rely on more devious and less reliable avenues of information. When I ask many historians of science why they feel that there is little commonality between their attempts to understand science and the efforts of their philosophical counterparts, or why they so rarely engage the larger questions about processes of scientific change and knowledge formation, the answers I get are quite diverse, though there are some common threads running through them. It is in reflecting on those answers that I have come to conclude that the reservations historians of science have about tackling philosophical (or other sorts of) theoretical issues about the dynamics of scientific knowledge rest on a set of dubious premises and ill-considered arguments. These arguments (which I shall attempt to summarise below) appear to function more as slogans for excusing non-engagement than as mature and carefully thought-through positions. It is some of those discussions which have prompted the remarks that follow.

Cognitive *vs.* Non-Cognitive History

One clear cause for the relative indifference of many recent historians of science to theories of scientific change has to do with shifting interests within the larger historical community about the central problems of history. Virtually all models of scientific change, whatever their differences, are concerned with explaining the vicissitudes of scientists' *beliefs* about what the world is made of. In a phrase, scientific theories and their cognitive fortunes are the central preoccupation of virtually all the philosophical theorists of scientific change. But, for a growing number of historians of science, there is no longer a consuming interest in understanding or explaining the cognitive fluctuations of the historical record. A concern with theory, or with the arguments and evidence adduced by scientists, smacks for some, these days, of a kind of outmoded 'internalism' or, even more offensively, of the study of 'disembodied ideas'. To many who take this line, the study of scientific ideas has been rendered obsolete by the emergence of the study of the social life of science. Science, we are told nowadays, usually in the portentous tones normally reserved for the enunciation of profound truths, is a human activity – nay, a social and human activity. Somehow the older focus on scientists' ideas is thought to de-humanise science by neglecting the larger social dimension in which science arises and by which it is transformed. According to some, it is not science, but *scientists,* which form the proper subject of historical narrative. These scientists seek to build careers, they organise themselves into societies, communities and disciplines. They jockey with one another for position and prestige, they form institutions, they found laboratories and especially if they are involved in 'big science' they build political careers and empires: they become scientist politicians. The new and apparently influential breed of social historians of science often attends to virtually everything about such scientists *except their ideas about the natural world*!

It is fairly clear why those engaged in history of science have little or no use for the available models of scientific change. Those models, after all, concern themselves primarily with describing how scientists come to form their beliefs about the world. Since the social historians are evidently uninterested in understanding the cognitive processes of the fixation of scientific belief, they see little promise in those models. I have no quarrels with institutional and more broadly social history when seen in its proper perspective. But I do believe that the questions addressed in such history of science are ultimately subordinate or peripheral to the main concerns of the history of science. Because social and institutional studies show some signs of coming to be regarded as harbingers or exemplars of a new genre of history of science, I want to explain why I find their concerns subordinate and secondary.

Science matters in our culture chiefly, albeit indirectly, because of what it says about the world. The political influence of science, its prestige, its level of financial support and even its size all depend on its theories and how those theories in turn have enabled scientists to anticipate and manipulate nature. Take away those theories, and the manipulative and predictive skills they confer, and science would arguably occupy a very different place in the intellectual landscape. To tell the history of science without explaining why scientists come to hold the beliefs about the world that they do is to confuse trappings with substance, effects with their causes. In sum, if it is true that science matters (both intellectually and institutionally) because of the manipulative and predictive skills which its ideas confer on their possessors, then a concern with science as a cognitive process must be primary, for until we have understood how science works cognitively, the largest question about science will remain unanswered. The theorists of scientific change recognise the centrality of the cognitive; that is why their theories focus primarily on the dynamics of scientific belief change.

Consider a close analogy. No one would entertain for a moment the idea that the central problem in the history of music was how symphonic performers organised themselves. They would reject out of hand the idea that the central problem in the history of painting should be the question of how painters' patronage was arranged. Yet it is today seriously being proposed (especially among English-speaking historians of science) that the *central* questions about science have to do, not with the master works which science has produced, but rather with the mundane minutiae of the life of a 'normal' scientist or with the ways in which scientists garner political support for their activities.

Such institutional historians of science, by failing even to confront let alone to answer the question of why scientists believe what they do, have opted out of the cognitivist game altogether. Many have evidently concluded that the only alternative to the disembodied history of scientific ideas is a lobotomised history of scientific institutions.

Resistance to Theory

Fortunately, however, many perhaps most historians of science are not committed to such institutional history as the sole or primary vehicle for understanding science. They seek to meet what I have just been calling the cognitivist challenge. If there is hope for constructive interaction between historians and philosophers of science, one might expect it to show up in the work of those concerned to explain the history of scientific ideas. Indeed, the cognitive historian of science is already dealing with epistemological issues, whether he cares to call them so or not. Typically, he is explaining why and how particular scientists discover, pursue, accept, modify and reject the theories of their time. What influence did Cambridge Platonism exert on Newtonian theories of space and time? How large a role did Darwin's observations on the *Beagle* play in shaping his notion of extinction? What importance, if any, did the Michelson-Morley results have for Einstein when he was developing the Special Theory of Relativity? These are precisely the sort of issues which modellers of scientific change engage. Yet the irony is that cognitively-orientated historians of science have shown little more interest than their institutionalist brethren in criticising, amending or even acknowledging the existence of the various theories of scientific change. As Kuhn once put it, 'few of the current practitioners of internal history are particularly philosophically minded'. Although virtually every piece of cognitivist history has implications for the general claims of theorists of scientific change, very few of the authors of such studies bother to explore their implications. This phenomenon is as perplexing as it is ubiquitous. What we are dealing with, after all, are distinguished historians, talking about particularly interesting and potentially revealing episodes of scientific change, who are wholly indifferent to exploring what implications their own narrative accounts have for a variety of widely discussed theories about scientific change.

It is rather as if a whole community of paleontologists collected enormous amounts of evidence about the extinction and distribution of species, but conspired never to reveal what their evidence entailed for prevailing theories of extinction and biogeography. I use the term 'conspired' deliberately; there appears to be an unstated rule of research in the history of science nowadays that one must never acknowledge the existence of any general theories about how science develops (not even to refute them), let alone use one's historical skills and evidential resources to go so far as to explore in detail how well or badly those theories fit the data.

When I ask my cognitively-inclined friends in history of science about the rationale for this self-effacing and self-denying convention, they often tell me that it is in keeping with the general tendency in historical research to eschew theorising or generalising from one case to another. The historian's duty, they say, is to tell a coherent story, preferably in narrative form, about a particular sequence of events. It is no part of the task of history, they insist, to subsume disparate and unique events under some over-arching pattern or grand design. As they see it, the construction or evaluation of 'theories' about the past is beyond the pale of legitimate historical inquiry. The authority of Herbert Butterfield and his repudiation of 'Whig history' is often invoked as a rationale. This is unfortunate since, although Butterfield was a fine historian, his published essays on historiography are a *melange* of confused *non sequiturs.* Yet it is Butterfield who seemed to sum up the anti-theoretical, anti-generalising bias of the current generation of historians of science when he wrote, in *The Whig Interpretation of History* (1931) that:

> *The eliciting of general truths or of propositions claiming universal validity is the one kind of consummation which it is beyond the competence of history to achieve.*

Because of Butterfield, the term 'Whiggism' has come to be used disparagingly of any general theory of scientific change, especially if that theory represents science as progressing through time. The Butterfield analysis, and that of all the other anti-Whigs who defer to it, appears to me to rest on several dubious assumptions about history in general and history of science in particular. Not the least of them is that it ignores the fact that several highly respected and influential branches of general history think of economic history, historical demography, labour history or much of social history engage quite explicitly in formulating and testing generalisations (whether they be about patterns of corporate behaviour, forms of economic life, the location of cities and ports, or the vicissitudes of crime and punishment). My first observation, then, is this: if it is legitimate to engage general theories in these well-established areas of history, why is it not so in the history of science?

Equally dubious is the assumption, implicit in the policy of non-engagement, that the writing of history can be, and ought to be, an atheoretical activity. One simply goes out, collects the relevant evidence, consults all the relevant archives and then proceeds ... how? To tell it as it really happened? Few historians would admit to being as naïve as that. But why is it any less naïve to imagine that theoretical views about the processes of scientific change can or should be cordoned off from writing the history of those changes? If the study of science itself has taught the rest of the world any lessons, it is that the most interesting research emerges from the interplay of theory and observation and that any effort sharply to separate the two is doomed. Yet, when it comes to their own craft, historians of science seem to exhibit the same suspicious resistance to engaging theoretical issues which was once associated in the natural sciences with narrow-minded Baconian inductivists. That whole approach to inquiry has been discredited, not least by the work of prominent historians of science.

Surely a degree of reflexivity is called for here. We know of few, if any, disciplines which, by limiting themselves to a narration of the 'bare facts' and by eschewing theory, manage to make significant and lasting contributions to the growth of knowledge. Whenever it has been tried, it has failed to produce significant results, for, try as one might, one cannot conduct research without bringing general presuppositions to bear. Why then does the historian of science imagine that virtue resides in writing about the history of scientific ideas as if no one had ever said anything of a general nature about how science develops? It is neither logically possible nor conceptually desirable to dissociate general and theoretical concerns from the construction of a narrative of specific events.

I have thus far argued for the logical impossibility of cleanly separating theory from practice in writing history of science. I have yet to address the desirability of intermingling the two, and that brings me to the real nub of the issue. The fact that general theories about scientific change have been developed, or even that they have achieved a fair degree of notoriety, is after all no compelling reason for the historian explicitly to engage them. (One might, however, have expected that the historian's much-admired passion for getting it right might have moved him at least to point out where these theories go astray.) I want to suggest that there is an urgent reason for taking this enterprise seriously, even if it should turn out – as I suspect it will – that most of the extant theories of scientific change will prove to be woefully inadequate. It has to do with the rationale or justification for history, and especially history of science, within the larger intellectual scheme of things.

Cast your minds back for a moment to that period in the 1950s when post war, English-speaking intellectuals were attempting to re-think the character of higher education. It had become clear that science and technology were key components of our cultural heritage and our contemporary life; it was equally clear that those activities had not received due scrutiny in academic circles. Indeed, virtually every major institution of our culture, whether the church, the military or the body politic, had received more systematic scrutiny from historians than had science. It was argued widely and cogently (*e.g.*, by James Conant, Vannevar Bush and C. P. Snow) that what was needed was serious scrutiny of science by scholars in the humanities and the social sciences. The point of that exercise was to make science *intelligible,* i.e. to explain something about how science operates, about the factors which impede or enhance its advance and about the relations between science and the other components of our culture. History of science and philosophy of science both profited institutionally from this thrust. Several countries set up funding panels for the humanistic study of science and universities founded departments, or more modestly, positions in history of science or philosophy of science where formerly there had been none. Degree requirements were rewritten so as to encourage students to study these disciplines in some depth. If it were not for that rewriting, most of the current practitioners of history or philosophy of science would be otherwise engaged.

The larger purpose behind these developments was not only to discover how science worked, but to communicate that understanding to scientists and the general public alike. History, quite properly, was given pride of place in this undertaking because it was thought that one could *learn* about science by scrutinising its past. That is why Conant published the *Harvard Case Studies* and that is why Kuhn wrote the *Structure,* a book which opens with the pregnant observation that:

> *History, if viewed as a repository for more than anecdote or chronology, could produce a decisive transformation in the image of science by which we are now possessed.*

But, as Kuhn would be the first to stress, history can teach lessons, and transform cultural images, only if it explicitly addresses *general* issues. A lesson, to be a lesson, must have general applications. If the episodes that the historian discusses have no significant elements in common, or nothing in common with the scientific experiences of our own time, then every episode becomes *sui generis* and we emerge from the exercise as impoverished as we were at the outset. Kuhn once chastised general historians for their 'abdication of responsibility for evaluating and portraying the role of science in the development of Western culture'. That indictment, thoroughly deserved, considering what short shrift science is given in general historical studies, is coming more and more to apply to historians of science as well. The latter, confronted with a choice between the Conant-Kuhn thesis that history can teach lessons, and the Butterfield doctrine that there are no general patterns to be gleaned from history, are moving dangerously close to refusing to evaluate the place of science in Western culture. I venture a conjecture: that virtually all historians believe that they have something to say to our time about how science works. But how are these insights to be communicated, and how are they to be authenticated, except by framing them as general theories or hypotheses?

Progress, Success and Whiggery Again

The divide separating many philosophers of scientific change from historians of science is broader still than is suggested by the distinction between those who make theories about cognitive processes of change and those who prefer to 'let the facts speak for themselves'. Philosophers' concerns are not only general and theoretical; they are also normative and evaluative. Philosophers of science generally believe that science has been a very successful enterprise for producing ideas which give us manipulative and predictive control over nature. They hold, moreover, that science through time has conferred greater and greater degrees of control over nature. In sum, philosophers generally see science as a *progress,* at least along this dimension, and they believe that a central task of historical research is both to record and to explain that progress.

Historians of science used to think so too. From Whewell to Sarton, traditional historians recorded the various steps on the progressive route of science from a largely ineffectual body of quasi-mythical philosophies of nature to the impressive achievements of a Newton, a Lagrange or a Maxwell. As late as the 1960s, historians of science still felt comfortable with these themes as guiding narrative threads. But, along with the rush from theory, historians have also been pulling back from an acknowledgement that science has become more successful through time and have dropped altogether any concern to explain the conditions which made that progress possible. Just as with the repudiation of theory, any concern with scientific progress as an explanatory challenge has been rejected as 'Whiggism'. As before, the brief has been developed at greatest length by Herbert Butterfield. He objected to any effort to write history as a progression because, as he saw it, that led to a falsification of the past; more specifically, it led to historians ignoring all those figures who did not lie on a straight line projecting from then to now. This, he thinks, is 'part and parcel of the Whig interpretation of history'.

One can sympathise with Butterfield's concern that a tale of victory, told only by the victors, makes for bad history. But in denying that historians are ever justified in recognising that certain parts of science are better than others, and in asserting that it is no part of the historian's task to explain the conditions which made them more successful, Butterfield (and those historians of science who follow him) would appear to be abandoning the programme of telling the full story of the past to which they are otherwise so deeply committed. It is, to take a specific example, a fact that Newtonian celestial mechanics is more successful empirically than Cartesian cosmology. It might explain much about science in general if we could come to understand what it was about Newton's way of proceeding which enabled him to produce a theory so strikingly more successful than its immediate predecessor. Philosophers, indeed, attempt to elicit just such an understanding. But if, with Butterfield, one denies at the outset that it is ever appropriate for the historian to ask whether one event represented progress over another, then one cannot even address the question, let alone answer it.

If he once grants that certain scientific theories or approaches have proved more successful empirically than their rivals, then the historian who disavows any interest in scientific progress is confessing that there are some facts about the past which he has no interest in explaining. Given that no history can be complete, that alone would not be very distressing. But the progressiveness of science appears to everyone *except the professional historian of science* to be the single most salient fact about the diachronic development of science. That, above all else, cries out for historical analysis and explanation.

If I seem to have dealt harshly with some of my historical colleagues, it is not because I see no promise in the historical study of science. On the contrary, I think historians of science have a splendid opportunity before them. Thanks largely to a concatenation of casual circumstances, there is now a large audience of scientists, philosophers, sociologists and others, looking to the history of science to move the discussion of how science works beyond the preliminary efforts of Kuhn, Feyerabend and the other early modellers of change. There are large intellectual issues which, it is now conceded, the history of science can uniquely illuminate. But if historians continue to refuse the challenge of giving a *general* account of scientific change and if they continue to imagine that science evolves altogether without direction and improvement, then others (especially philosophers and sociologists), possibly less well-suited for the task, will step into the breach. Indeed, they are already doing so. Kuhn, for one, is on record as predicting that this process is inevitable, for he has lately questioned whether 'professional' historians of science are any longer prepared to write the history of scientific ideas with a view to addressing questions of a philosophical or theoretical sort.

4

Beginning Era of Science and Technology

Introduction

The Paleolithic period was a preliterate epoch which lasted from roughly two million to 10,000 B.C., while the Neolithic age was a later preliterate period that lasted from about 9000 B.C. until the first civilizations of Egypt and Mesopotamia. The people of the preliterate eras may have even classified a variety of plants and animals, although in no sense did they have any knowledge of these facts as a "science." However, the shrewdest of the early species of humans of the later Paleolithic era, Cro-Magnon, at the very least, would have soon found out which plants were toxic, which were not, and which were suitable for medicinal purposes. They also probably knew which kinds were appropriate for dyes, poisons, and so forth. Nor did organized biology begin to emerge even with the earliest true civilizations in the West the ancient empires of Egypt and Mesopotamia, which began roughly 3500 B.C. Despite considerable progress in the arts and culture, generally in Egypt and Mesopotamia and the many civilizations that followed them, neither science per se nor organized biology really existed as a separate, organized body of thought. This would remain so at least until the time of Aristotle, in the fourth century B.C. In the case of biology specifically, whatever knowledge did emerge would remain intertwined with medical practice, again, until the time of Aristotle. The biological knowledge the ancients did discover and develop came about primarily because of practical surgical and medical needs between the time of ancient Egypt and Aristotle, frequently out of a necessity to aid soldiers in the battlefield.

Egyptian Medicine

Thus, at least the rudiments of biological/medical learning undeniably existed in the Egyptian age. Five thousand years ago, Egyptian priests were already starting to gather a tremendous amount of medical data. We know this from their hieroglyphic stone tablets, many of which linguists have almost completely deciphered. Among other things, they ostensibly had a highly advanced knowledge of plants and their various medical applications. Thus, the real story of Western biology begins in Egypt.

The reason for this is not hard to understand. The study of ancient civilizations has shown that cultural progress cannot take place in periods of rampant misery. Progress in philosophy, abstract science for its own sake, literature, art, and the like is greatest when life is most serene and unburdened with problems. There are innumerable examples: one thinks of Greece in its "golden age," where men like Pericles and Cleisthenes ruled democratically in a nation that offered almost anything one could want. In such an atmosphere, literature and philosophy flourished through people such as Aristophanes, Plato, and Aristotle. It is no surprise that in the work of Aristotle we do in fact see the study of biology for its own sake. Aristotle was interested in medicine, but he also conducted myriad scientific investigations merely for the sake of learning about what he did not understand, unburdened with concerns for feasible "practical" applications.

Unfortunately, in Egypt, there were few extended periods of peace or tranquility and, therefore, little leisure time for the pursuit of abstract science. (Historians generally divide the history of Ancient Egypt into the Old, Middle, and New Kingdoms, with the Old Kingdom beginning at about 2770 B.C., and the New Kingdom ending about 1087 B.C.) Unquestionably, this was true of the Old Kingdom, where one sees such tyrants as Zoser, who spent virtually all of the public money on a grandiose and wasteful series of shrines to himself called the pyramids. Perhaps the closest Egypt came to a "golden age" of unrestricted science and art was during the Middle Kingdom, which lasted from about 2000 to 1785 B.C., and which comprised the eleventh and twelfth dynasties. In this period, the pharaoh had more or less made peace with the nobles and priests who had challenged the might of the earlier Pharaohs. Also, with the advent of fewer egomaniacal Pharaohs, the frenzy of pyramid building had begun to diminish. In turn, there was considerable progress in astronomy, math, and the practice of medicine. Mathematics, including geometry and arithmetic, developed to aid in massive construction projects, and astronomy developed to compute the times of the flooding of the Nile. In medicine, various documents dating from about 1700 B.C. show that both diagnosis and treatment of disease were highly developed. The Egyptians had learned the significance of the pulse and the circulatory system, could treat broken bones, and, as the ancient *Materia Medica shows*, had catalogued quite a list of medicines.

In the late Middle Ages, by contrast, life was one continual, dreary struggle for the next meal. Life was filled with constant plague, and people had little hope for a better life. It is thus no wonder that philosophical, scientific, and literary accomplishments were comparatively few in the later medieval period. There were exceptions of course: the fourteenth-century philosopher William of Occam is, defensibly, one of the exalted thinkers of antiquity. But, in general, this law of history holds true.

One of the earliest "physicians" then of ancient Egypt was Imhotep, who apparently lived sometime during the third dynasty, or first ruling family of the Old Kingdom. A versatile man, historians believe he also orchestrated the construction of the first pyramids under Zoser, one of the earliest of the legendary Pharaohs. There is some reason to believe, based on what has been learned from written records, that early Egyptian physicians like Imhotep acquired a respectable amount of their medical information from the custom of mummification. The first step in this process consisted of carving out the innards of the deceased. This, of course, gradually allowed early physicians to accumulate considerable anatomical data. Then, by experimenting with different plants for preserving the internal organs, they acquired some knowledge of biochemical action. By trial-and error testing of a variety of chemicals on the body to determine their biological effects, they found which ones preserved the body best. In this group were wine, aromatics, myrrh, cassia (a legume with medical properties), salt, and several other chemicals. They realized too that extracts from a wide selection of plants covering the linen wrapped body would also inhibit decay.

Myth and Medicine

Nevertheless, it was nearly impossible to separate magic and superstition from science. In part this was a function of ignorance, in part a function of power. Throughout the Old, Middle, and New Kingdoms, the Pharaohs had to contend with the domination of the priests. Possibly because the masses believed them to have magical powers and influence over the deities of nature, the priests had immense leverage, and no Pharaoh could afford to ignore them. Imhotep was among the most commanding and influential of the priests. Many centuries later, by the time of the New Kingdom, many regarded him as essentially a divine being. The masses worshiped him, offered gifts to bronze images of him, and prayed to him to let their crops grow.

Medical Papyri

A measure of how far the ancients were from modern science is the fact that they did all of their research under the belief that spirits dwelt in terrestrial objects. The Egyptians believed that some god or other was the steward of every part of the body. However, specific papyri, such as the famous *Edwin Smith Surgical Papyrus*, show that not all of their medicine was perceived as a mere appendage of religion. For example, during the New Kingdom, which was the last of the major epochs in Egyptian history, surgery, according to the papyrus, was free of theology. The Egyptians judged that purely mechanical damage to bones, muscles, and joints had natural, rather than theological, explanations. Even so, much of the Egyptian population still believed that cures came only through the worship of divinities such as Aton, the lord of the universe in the New Kingdom, Amen-Re, the sun god, and others.

The same papyrus also explains that by the period of the New Kingdom, Egyptian physicians understood how to set bones very well. Even their diagnostic procedures resembled contemporary medical practice; in the Edwin Smith papyrus there are various lists of questions, such as "Can he lift his arm?" "How rapid is his heartbeat?" and so forth. Although the papyrus is not entirely explicit on the reasons for these questions, it seems a safe bet to assume that they were diagnostic. The papyrus also discusses injuries to many different parts of the body, including the face, head, and abdomen. It tells how to bandage battle-ax as well as knife wounds, and how to apply tourniquets.

The Egyptians knew that the heart was a pump of some sort, and they may even have had a rudimentary comprehension of the circulation of the blood. Interestingly, the Greek pre-Socratic philosopher Empedocles also studied the heart and circulatory system, though today historians know him best for his philosophical views. Although he made many errors in his descriptions, such as claiming that the human psyche was located in the heart, his first stumbling forays were important stimuli to later research on the circulation of the blood.

The Egyptians acquired all of their erudition through the method of trial and error. We know, for instance, that they first realized the antibacterial properties of penicillin through this method. They had no theory as such; they knew only that certain techniques helped particular ailments. In like fashion, they came to know that onions were valuable in protecting them against epidemics, apparently functioning as a kind of primitive antibiotic, and that the yarrow plant, by constricting blood vessels, functioned as does a modern styptic pencil to control bleeding. Finally, although they had insufficient knowledge of alcohol's properties, they shortly discovered that giving alcohol to patients, especially via mixing beer with other medications, gave the patient a decidedly useful sense of well-being. The Egyptians also acquired some biological knowledge through agriculture. Even in preliterate ages, as early as 8000 B.C., humans were learning about the soil conditions necessary for raising wheat and corn. In doing so, they acquired practical botanical knowledge about how soil conditions such as moisture, acidity, and the like affected the development of plants.

Papyrus Ebers

Another well-known text, the *Papyrus Ebers*, offers descriptions of over 600 medicines, explaining in considerable detail how the

Egyptians used each of them. Allegedly, it was over a foot wide, over sixty feet long, and written wholly in hieroglyphics. According to legend, a nomad discovered it near the Nile River sometime during the Middle Kingdom of Egypt, although assorted historians believe it was written in about 1550 B.C., or around the end of the second intermediate period. The sections of it that have survived rest permanently today in Leipzig, Germany. This historic papyrus prescribes exercise, hypnosis, diet, and fasts. More specifically, it discusses the function and dysfunction of the various senses and their sense organs, including long sections on ailments of the ear and eye. It says much about tumors and blocked glands and offers various medicines designed to heal burns, wrinkles, freckles, and nearly every skin condition imaginable. It prescribes medicines for baldness as well. Some sections deal with pediatrics, for medical specialization was common in Egypt.

Therapeutic Papyrus

The Therapeutic Papyrus also suggests well over 600 distinct remedies for a broad spectrum of ailments. The remedies include copper, acacia, castor beans, olive extract, saffron, and ginger. Now and then the "medicines" border on the preposterous. For example, parts of this papyrus imply that moisture from a sow can cure illness. Furthermore, the information in the papyrus was not purely medical. It is evident from the *Therapeutic Papyrus* that Egyptian biologists knew something about how tadpoles changed into frogs and how beetles emerged from eggs.

Many Egyptian cave paintings as well as hieroglyphics show explicitly enough that the Egyptian priests were practicing surgery of all types. They apparently did abdominal surgery, surgery of the groin area, and even eye surgery much of it successful. It is arguable that there was not a comparable achievement in surgery until the mid seventeenth century, when the Germanic physician Johann Shultes wrote his *Armamentarium Chirugicum*, or "Tools of the Surgeon," which proposed, among other things, an unpleasant procedure for performing a mastectomy. Despite their superstitious beliefs concerning science, Egyptian approaches at times were strikingly reminiscent of recent medicine.

5

Language, Discourse and Science

Introduction

Language has emerged as a central concern of many areas of the humanities and social sciences in recent decades. The increasing pervasiveness of the highly ambiguous term 'discourse' in much contemporary debate is a result of this concern, which has so far extended over the disciplines of philosophy, anthropology, sociology, and literary criticism, and is beginning to have an effect in history itself. To date most historians of science have proved reluctant to execute a 'linguistic turn', and to join their colleagues in other fields in giving serious attention to problems of language. Their reluctance is regrettable, since it passes up the opportunities that linguistically-orientated studies would offer of forging new connections between the history of science and other contemporary disciplines; but it is also perhaps understandable in view of the disagreements and confusions which exist in the current understanding of language. For, despite widespread agreement that language is the central problem of modern human studies, there is no unanimity as to what this implies methodologically. As Martin Jay has put it, 'Linguistic turns may take many different directions.'

Beneath the apparent coherence of the modern attack on the problem of language lie widely disparate philosophical views on its nature and human significance. Thus, while most analytical philosophers would regard language as an entity which mediates between the human subject and objects in the external world, others such as structuralists and post-structuralists see it as an autonomous system governing the construction of meaning independently of human intention or reference to the objective world. To some, language is a defining feature of human existence, to others (such as semioticians, structuralists and hermeneuticists) it provides a model for the understanding of all human action. In view of this diversity of available philosophical perspectives on language, the urgent need is for some clarification of the theoretical issues at stake, and discussion of the ways in which they have been and might be applied to research in the history of science. Brief as it must be, this essay will attempt to address these needs.

That the turn towards the study of language is deeply rooted in the fundamental assumptions of many areas of contemporary human studies can hardly be denied, though it has never been satisfactorily explained. Michel Foucault has however suggested elements of an explanation in his brilliant but much-contested 'archaeology' of the human sciences, *The Order of Things*. Foucault argues that language did not exist as a problematic entity in the eighteenth century, in the sense that it was regarded simply as a means of representation, albeit one which was not always properly used. With the rise of historical studies of grammar in the early nineteenth century, language lost its transparency, and acquired a historical 'density'. It became an attribute of the newly-constructed entity, 'man'; it was both an object of knowledge and also an irreducible limitation on what could be known. This development had two consequences, in Foucault's view. Firstly it stimulated consistent attempts to construct a formal, logical representation of thought, which would be free from the ambiguities and loose nesses of human language. Secondly, there developed a critical focus on the various ways in which language functioned as an *a priori* determinant of what human beings could think. Marx, Freud and Nietzsche are portrayed by Foucault as contributors to this latter enterprise, which has revealed how much we are all 'governed and paralysed by language.'

In this way, Foucault suggests that two apparently opposite aims, the reduction of human thought to its logical form, and the interpretative understanding of the way in which language constrains thought, are both in fact symptoms of an underlying awareness of the density of language, an awareness which is characteristic of the modern view of man. Similarly, Ian Hacking has noted how several apparently distinct philosophical traditions of our time share what he calls 'immense consciousness of language'. Foucault's point is also paralleled by that of Richard Rorty, in a review of the evolution of recent British and American linguistic philosophy. Rorty claims that both those philosophers who sought to solve philosophical problems by a logical formalisation of language, and those who looked for solutions through a closer understanding of the common usage of language, shared the basic belief that a philosphical problem was 'a product of the unconscious adoption of assumptions built into the vocabulary in which the problem was stated'.

The identification of this basic assumption of much recent philosophy, and of many other areas of the human sciences, provides one reason for historians of science also to take language more seriously. To take the 'linguistic turn' in this sense is to relinquish the view that the language of science (any more than that of philosophy) is simply a means of representation, separable from and irrelevant to the 'content' of science. Such a move carries with it the implication that historians should no longer seek to penetrate *through* the linguistic practices of scientists, to isolate ideas and conceptual structures in their minds, and should instead start to investigate those verbal and textual practices themselves.

Historians who take this step are undertaking to look sensitively at the ways in which scientists use language in writing texts, giving lectures, preparing research reports, holding conversations, and so on. I shall adopt that embarrassingly overloaded term 'discourse' to cover all of these practices. But to adopt the term is not to specify how it should be used; and to take the 'linguistic turn' is not to commit oneself to any particular mode of analysis of scientific language. There remain in fact a number of options available. In the discussion that follows, some examples of recent historical research on scientific language will be mentioned, and related in each case to the philosophical view of language that appears to underlie the investigation. The purpose of such a discussion is to clarify the theoretical issues in the field, not to suggest that historians are (or should be) guided by the views of particular philosophers. For the sake of simplicity, I shall distinguish just three kinds of approaches, characterised by three distinct attitudes to language: the 'symbolic', the 'hermeneutic or semantic' and the 'rhetorical'. Broadly, the first approach studies language in relation to the objects to which it refers, the second interprets it as a system of meanings, and the third analyses the way in which it serves the interests of its producer. These categories are intended to aid analysis, rather than to restrict it; it is not suggested that all relevant work will fit neatly within any one of them.

'Symbolic' Approaches

The 'symbolic' approach takes as its object of study the ways in which language exercises its function as a system of symbols. It starts from the assumption that language is a means of symbolic communication, that it attaches verbal signs to things, and uses them to convey meaning. Rather than seeking to grasp the meanings that the signs are given, this approach concentrates on understanding the way in which the connection between sign and thing is made. The connection is studied in the light of views on the nature of language articulated by those whose linguistic usage is under investigation, or by others in their cultural milieu.

The rationale of this approach has been described by Ernst Cassirer. In his work, *The Philosophy of Symbolic Forms,* Cassirer portrays language as the primary attribute of human culture. But he makes it clear that he wishes to grasp the function of language symbolically, rather than semantically. Philosophy, he claims, cannot penetrate behind language; it can only discern the way in which language works, namely by attaching labels to sense-data, so that objects are brought within the range of the understanding. Linguistic symbols denote objects, and thereby appropriate them within the structure of the mind through which the world is apprehended.

On this view of language, the interesting questions about linguistic usage concern the way in which symbols are attached to the objects they name or denote. Two overlapping pairs of alternatives must be considered. Firstly whether the denotation is made unambiguously, *i.e.,* whether there is a one-to-one correspondence of word with thing; and secondly, whether the naming is taken to be entirely conventional, dependent solely on human choice, or is thought to rely on some essential (natural or divinely-revealed) connection between an object and its name. Several historians have investigated shifts between these alternative methods of denotation, particularly in the crucial period of the seventeenth-century Scientific Revolution, and have related them to concurrent changes in scientific and philosophical ideas.

On the issue of ambiguity, the options are distributed between two extremes: at one pole, unambiguous word-for-object denotation; at the other, complete freedom in attaching any name to any thing. Although in fact neither extreme is possible, all language occurring at some point between the two, there is pressure in some contexts for denotation to be more precise and ambiguity to be restricted. For example, historians of the seventeenth century have portrayed the 'plain style' espoused by some natural philosophers of the time as an attempt to push usage towards the pole of unambiguous naming. Thomas Sprat, a prominent apologist for the new science as practised by the Royal Society of London, denounced 'this vicious abundance of *Phrase,* this trick of *Metaphors,* this volubility of *Tongue,* which makes so great a noise in the World', and called for 'a close, naked, natural way of speaking'. John Locke's critique of linguistic inaccuracies, his identification of six types of 'abuses of words', and his castigation of rhetoric as 'that powerful instrument of error and deceit', developed and formalised themes which Sprat and other seventeenth-century natural philosophers had already explored.

Schemes for artificially constructed 'universal' or 'philosophical' languages, which were also commonplace in the seventeenth century, appear to have taken this desire for unequivocal denotation to its practical conclusion. In these schemes, precisely one name was to be given to each thing, and compounded objects were to be given names synthesised from the names of their parts, so that the processes of analytic and synthetic reasoning could be exactly reproduced linguistically. Historians such as James Knowlson, Vivian Salmon, Hans Aarsleff and Mary Slaughter have studied these proposals, found in the writings of J. A. Comenius, George Dalgarno, Francis Lodowyck and John Wilkins, among others. They have pointed out various connections with scientific developments of the period, but have not reached any consensus as to what the original motivation for the schemes was.

In Aarsleff's account, the issue of conventionality is crucial in this connection. In his view, the schemes for 'universal languages' indicate the survival of the 'Adamic' view of language, which holds that words and things were originally connected by God's command at the Creation. Although the proposed languages were to be explicitly artificial constructions, their espousal demonstrates to Aarsleff that it was still thought possible to recover the original divine link between language and the world. The 'Adamic' view of language had its roots in pietistic mysticism, feeding through from Protestant Germany to the English Puritan reformers of the mid-seventeenth century. It was displaced towards the end of the century by the view, expounded forcefully by Locke, of the inescapably conventional nature of language as an entirely human construction.

To Slaughter, on the other hand, the universal language schemes bespeak the continued predominance, until the late seventeenth century, of what she describes as a 'taxonomic episteme', based on Aristotle's philosophy of essentialism. They mark the final stages of a basically scholastic enterprise of exhaustive classification of natural objects, with the aim of knowing them in their essences. In this way Slaughter differs from Aarsleff on the real motivations for the universal language schemes, though she agrees on the chronology of their rise and decline, and on the importance of Locke in sealing their eventual fate. On balance, it seems likely that both Aristotelian essentialism and the Adamic language doctrine were influential upon those who proposed linguistic reforms in the seventeenth century, though neither seems established as a completely comprehensive explanation of the popularity of such schemes.

While the historians remain divided over significant details in this area of scholarship, the basic orientation of their studies seems clear. Their work falls under the heading of what I have called the 'symbolic' approach, because they assume that the symbolic function of language, the way in which names are attached to objects, is the relevant issue. They thus devote attention to the practices of naming, whether ambiguous or precise, conventional or natural. The assumption made is that the choice of particular methods of denotation on behalf on language-users is a voluntary one. Hence, expressed views about the nature of language can be used as evidence of usage, and connections between, for example, scientific use of language and philosophical views *about* language in the relevant milieu, can be expected to be uncovered. Geoffrey Cantor's research on connections between theories of language and theories of light in the eighteenth century illustrates how this approach can be extended into the history of science in the period since the Scientific Revolution.

'Hermeneutic or Semantic' Approaches

The second category of approaches to scientific language is the one I have called 'hermeneutic or semantic'. I am applying this label to studies which give serious attention to the problems of interpretation of the meaning of scientific language, in other words, which do not follow Cassirer in restricting their attention to the mechanisms of denotation. In a sense, of course, historians have always been concerned with meaning; some measure of interpretation in inseparably part of their enterprise. But when they take the linguistic turn, they are forced to confront afresh the problems of interpretation of past scientific theories. Meaning is now seen, in science as in other linguistic activities, as emerging from an interaction between the human subject, the world, and the formal structures of the language in which it is expressed. In future, history of science which aims at the interpretation of meaning will have to take account of the wide range of linguistic codes in which the substance of science has been communicated.

Recent historical research has used this kind of approach to illuminate two particular types of problems. Firstly, the problem of the meanings given to scientific theories in certain cultural contexts. Studies in this area offer new perspectives on the popularisation of science. Secondly, the question of metaphor in scientific language is also illuminated by this kind of study. Whereas in the 'symbolic' approach, metaphorical language was seen simply as an imprecise method of denotation, a 'hermeneutic' approach permits metaphors to be read as devices for the transfer of meaning between different disciplines, or between science and general culture.

A brilliantly-sustained recent analysis of the first kind of problem is Roger Cooter's study of nineteenth-century phrenology, *The Cultural Meaning of Popular Science.* Cooter cuts across simplistic accounts of the popularity of phrenology at this time with a detailed analysis of the meanings ascribed to phrenological discourse in various cultural contexts. This involves him in discussion of the ways in which written texts were interpreted, the iconography of illustrations, and the messages conveyed by popular phrenological lectures. In each case, the historian is interested in processes of interpretation and the construction of meaning, and at the same time his own procedure is explicitly interpretative, in seeking to capture the meanings of past discourse. Thus Cooter seeks to grasp the appeal of phrenology to its audiences 'by decoding the science's signs and symbols'; he exhibits those symbols as 'personal, intimate, common, flexible, and "real" (because of their external palpably empirical nature)', and describes in great detail the variety of personal meanings which their flexibility might make them able to carry. Phrenology emerges as 'a multifaceted symbolic resource'; a mode of discourse whose success stemmed from its extraordinary adaptability as a way of making sense of a wide range of human social relations.

This brief summary does no justice to the complexity of Cooter 's analysis and it neglects altogether his argument for the importance of the symbolic functions of phrenology in relation to the stability of Victorian society at large. My point is to show the potential productivity of an approach which attempts to interpret the meanings of scientific discourse in specific cultural contexts. Such a perspective allows historians to dispense with the idea that science is always straightforwardly and inevitably 'popularised', and encourages us to grapple with the problems of understanding the actual meanings which scientific discourse had to those who comprised its audience.

This perspective is also capable of yielding new insight into the importance of metaphorical language in science. Semantically, metaphor can be understood as a transfer of a term between different contexts of meaning. Metaphors thus emerge, in the 'hermeneutic' approach, as crucial for understanding conceptual transfer between scientific disciplines, or between science and other areas of culture.

A stimulating discussion of this has been provided in a paper by Karl Figlio on the metaphor of organisation in the early-nineteenth-century bio-medical sciences. Figlio describes metaphorical language in science as a link between the formal elements of scientific theories and the ideological role of science in general culture. His discussion touches on the example of notions of irritability in eighteenth-century physiology, notions which while 'precisely specified in a technical sense, were obviously infused with all the shades of meaning attached to ideas of life'. The wide metaphorical implications of the terms employed allowed for the investment of crucial social interests in debates in this area. Questions of materialism, determinism, and others, were hotly contested; and Figlio holds that this was because a range of important personal interests were involved in commitments to one side or another. He concludes that an understanding of the metaphorical meanings of scientific terms provides a method by which these social investments in science may be exposed:

> *It remains the task of the history of science to elucidate these interests philosophical, theological, social, national, or political (i.e. ideological) but it is an unapproachable problem until the concealed level at which these interests impinge upon each other is uncovered; i.e. until the meanings conveyed in the scientific language in which they are expressed are unveiled at the level at which ideas metaphorically flow between apparently separate domains.*

Thus the interpretation of the role of metaphor in scientific language appears to be essential, if historians are to understand the way human interests are invested in science. A study which parallels Figlio's in this respect is Robert M. Young's discussion of Darwin's metaphor of selection. Young shows that the term 'selection', as used by Darwin, functioned technically in a description of the mechanism of evolution, but that it also had wider connotations which included an evocation of divine design. The term therefore served to weld together a diverse group of adherents to Darwin's theory, covering a wide range of different views as to the extent of God's intervention in the world.

The 'hermeneutic or semantic' approach which I have attempted to identify thus offers promising new perspectives on the relations between science and general culture. By undertaking the interpretation of the language of science, it offers the prospects of grasping the meanings of scientific terms within and outside technical disciplines. It is nevertheless largely unperceived as a distinctive approach in the history of science; and this is a pity, because the price of non-recognition is that the methodological problems of interpretation are not confronted explicitly, and the possible links with other disciplines are not explored.

Other groups of historians have been bolder in advancing this kind of method. For example, both Keith Michael Baker and William J. Bouwsma have made the call for intellectual history to be transmuted into 'the history of meaning', a history which would embrace all aspects of culture and all forms of symbolic expression of human experience. Although the history of science would presumably wish to secure the boundaries of its own territory within such an all-embracing 'history of meaning', an expanded awareness of the range of human symbolic activity, and of the methods of its interpretation, would surely offer new views of a wide range of issues concerning the relations between science and human culture at large.

'Rhetorical' Approaches

The third category of methods for studying scientific language, the 'rhetorical' approach, concerns itself not with the meaning of language, but with its other functions. In its widest sense, the discipline of rhetoric studies the ways in which language is used in argument, the means by which persuasion is effected by a certain form of words deployed in a specific relationship between speaker and audience (or, by extension, between writer and readership). In other words, rhetoric deals with the uses of language, as opposed to its meaning; or with its ritualised production in certain situations, as opposed to its conceptual content.

Unfortunately, philosophical discussions of rhetorical theory have been few and far between. The ancient antagonism between philosophy and rhetoric, dating from Socrates' arguments with the Sophists, has inclined philosophers to believe that rhetoric is unworthy of their attention, except in so far as it requires denunciation in the terms which Locke pioneered. On the other hand, some recent theorists of language, including hermeneuticists and semioticians, have been keen to link their approaches to that of the ancient discipline. And historians of science have also been given increasing attention to rhetoric in recent years.

One aspect of this has been the recovery of the historical importance of the discipline, particularly in relation to the Scientific Revolution; another has been the attempt to discern rhetoric in action within scientific discourse of various forms and periods. The second of these fields will be our main concern here, although it is worth noting that significant work has also been done in the former area. From the research of Paul Oskar Kristeller since the 1940s, it has been recognised that the revival of classical rhetoric was an important part of the learned culture of the European Renaissance. Historians such as Walter Ong and Lisa Jardine have mapped changing attitudes in the Humanist disciplines of rhetoric and dialectic, and have suggested how such changes might have had a lasting influence on the progress of the Scientific Revolution. More recently, scholars such as Steven Shapin and Peter Dear have been showing how scientists of the seventeenth century faced problems concerning the literary representation of their experiments, and the mobilisation of linguistic resources to persuade their readers; problems which were implicitly rhetorical. Barbara Shapiro has further argued that the methods and standards of demonstration which were found appropriate for deployment in the new science were those of traditional rhetorical fields, such as the law. From this work, an outline picture is emerging of the long-term importance of the renaissance revival of rhetoric, as a contributory factor in the seventeenth-century Scientific Revolution.

Quite independently of the results of this historical enquiry into the influence of rhetoric as a discipline, it can be argued on theoretical grounds that rhetorical properties always inhere in scientific language. Any use of language to persuade or move an audience can be analysed rhetorically, and this applies to science as to all fields in which language is used. The distinctive features of such a rhetorical analysis are twofold. Firstly, it does not distinguish between the 'content' of science (its theoretical ideas) and the 'style' in which they are presented. Rhetoric has always treated content and style as inextricable; its analysis is focused on the ritual regularities of expression which make a particular piece of language representative of a certain oratorical or textual genre. It is this feature of rhetorical analysis which gives it appeal to modern linguistic theorists anxious to deny 'metaphysical' distinctions between content and style in discourse. Secondly, rhetorical analysis concerns itself with the relationship between speaker or writer and audience, and in this way opens out to wider historical study of the contexts of scientific communication and the functions of scientific texts.

Understood in this sense, an analysis of scientific language as rhetoric can find encouragement in the writings of authors as widely influential, but as divergent in their general outlook, as Thomas Kuhn and Michel Foucault. One of the implications of Kuhn's crucial concept of a 'paradigm', introduced in his major work *The Structure of Scientific Revolutions*, is that some kind of rhetorical analysis of science is called for. Kuhn describes a paradigm as a model problem-solution, concretised in the form of a textbook and perpetuated by a system of instruction, from which it cannot be entirely abstracted. As David Hollinger has indicated, Kuhn's importance for historians lies primarily in his having related scientific practice to the continuing authority of tradition; a tradition which is understood not as comprising certain abstract principles or concepts, but as embodied in the textual forms of concrete exemplars enforced as conventions of practice. To adopt M. D. King's phrase, Kuhn has made science out to be, not a world view or a set of methodological prescriptions, but 'a system of traditional authority'.

If he is interpreted in this way, Kuhn's analysis has points of contact with that of Foucault in his 1971 address, 'L'Ordre du discours'. Foucault commends a study of discourse which would concentrate particularly on the institutions of pedagogy, such as academic disciplines, which 'constitute a system of control in the production of discourse'. He therefore urges that we are 'not to burrow to the hidden core of discourse, to the heart of the thought or meaning manifested in it; instead, taking the discourse itself, its appearance and its regularity, we should look for its external conditions of existence'.

Given the recommendations of Kuhn and Foucault, it is perhaps unsurprising that one of the main problems recently illuminated by a rhetorical perspective on scientific language is that of the maintenance of disciplines. Wilda Anderson, for example, has suggestively discussed the contribution of certain rhetorical forms to sustaining the identity of the discipline of chemistry during the Enlightenment. Focusing on the *Dictionnaire de Chymie* of Pierre-Joseph Macquer, she shows how its structure manifested the possible interconnections of the different parts of chemistry, but avoided imposing a dogmatic system upon it. Macquer's introductory review of the history of the subject also consolidated its unity by providing a historical legitimacy for its practitioners, and justified its claims to be a 'philosophical' discipline with authority to oversee a variety of practical arts.

Other historians have applied a rhetorical approach to understanding the functions of public addresses and orations. These have been shown to utilise traditional linguistic forms to appeal to certain audiences, thereby securing the legitimacy of disciplines and the positions of particular practitioners therein. Herman Boerhaave's orations at the University of Leyden in the early eighteenth century, the eulogies of Vicq d'Azyr in the Societe Royale de Medecine in the 1770s and 1780s, and the *eloges* of Georges Cuvier in the Institut de France in the 1800s have all been subjected to rhetorical analysis of this kind.

Aside from serving to sustain the unity of disciplines, rhetoric in scientific language can have other functions. For example, the means by which scientific writings attempt to convey a representation of the 'facts' of experiments can also be seen as inherently rhetorical. A number of writers have advanced the argument that scientific language can never be simply and purely descriptive, that it inevitably deploys linguistic artifice and formal devices to persuade the reader of the 'factual' status of what is described. This point of view has been developed, and applied to examples of modern scientific research papers, by Joseph Gusfield, Steve Woolgar, Nigel Gilbert and Michael Mulkay, among others.

Some of these writers have argued that, because scientific 'facts' are rhetorically constructed, analysts should abandon any attempt to read through scientific texts to the social or experimental context in which they were produced. Gilbert and Mulkay, for example, argue that any sociological account of scientists' practices of operating upon nature would be an arbitrary imposition of a single meaning upon the many varieties of scientific rhetoric. The sociologist's is just one account among many, and 'almost every single account is rendered doubtful by its apparent inconsistency with other, equally plausible, versions of events'.

To this kind of argument, historians are likely to respond that the variation among possible accounts of science does not by itself indicate that choice between them is entirely arbitrary; it simply raises the question of the methods which are appropriate to elucidate a valid historical description. Arguing along these lines in their recent book, *Leviathan and the Air-pump,* Steven Shapin and Simon Schaffer have used techniques of the rhetorical analysis of scientific discourse, but have not shirked the historian's duty to place that discourse in its historical context.

Shapin and Schaffer have investigated the pioneering construction of a language of experimental description by Robert Boyle in the mid-seventeenth century. They describe how every feature of what they call Boyle's 'literary technology' was tailored to the purposes of conveying a persuasive account of the experiments he described. The prolixity of Boyle's prose, his relentless description of circumstantial details, even of failed experiments, and the meticulous diagrams of apparatus he provided, were all means towards a basically rhetorical end: the persuasion of the reader of the factual status of his accounts. 'Appropriate moral postures' also enhanced the persuasiveness of Boyle's descriptions; these included candidness about failures, modesty in advancing opinions and a reluctance to engage in disputes. And in addition, the factual credentials of the accounts depended upon their being correctly witnessed by reliable men, preferably of gentle birth, and on the names and rank of those witnesses being recorded. According to these authors, many of the features of the rhetorical construction of scientific research papers, which sociologists of science have identified in modern examples of the genre, had their origins in the verbose writings of Robert Boyle.

Shapin and Schaffer push their analysis further, beyond the level of the rhetorical devices in Boyle's texts. By exploring the differences between Boyle and Thomas Hobbes, they show how disagreements about the interpretation of experiments disclose fundamental epistemological differences over the way in which knowledge was to be constructed, and these in their turn are connected with alternative political visions. The social order or 'polity' which was constructed among experimentalists, or between them and their audiences, was integrated within the social and political order of society at large. In this way, Shapin and Schaffer link the experimental and discursive practices of Robert Boyle and his allies in the early Royal Society with the consolidation of social and political stability in Restoration England in the 1660s.

All three approaches seem compatible with the historian's usual aim of understanding the development of science in its cultural and social context; they thus promise to reward further development and application in historical research. There remain of course theoretical problems, which I have not been able to broach in an essay of this length. This review is intended to point the way to such further investigations, by suggesting a pattern for understanding what has already been done.

6

Marxism and the History of Science

Introduction

The main thing to be said about Marxism and the history of science is that more, much more, has been written that is explicitly or implicitly anti-Marxist than has been written which is avowedly an attempt to see the history of science in Marxist terms. Another class of writings can be seen as watered-down Marxism, while still another is silent about Marxism but does not make much sense unless one knows that Marxism is the silent partner in a one-voice dialogue or polemic in which the other position is not named. The analogy which springs to mind is that of a planet which is not seen but is inferred because of the perturbations of the other planets due to the gravitational effect of the unseen one.

The defining feature of Marxist approaches to the history of science is that the history of scientific ideas, of research priorities, of concepts of nature and of the parameters of discoveries are all rooted in historical forces which are, in the last instance, socio-economic. There are variations in how literally this is taken and various Marxist-inspired and Marxist-related positions define the interrelations among science and other historical forces more or less loosely. There is a continuum of positions. The most orthodox provides one-to-one correlations between the socio-economic base and the intellectual superstructure. This is referred to as economism or vulgar Marxism. The classical source is a set of comments in the preface to *A Contribution to the Critique of Political Economy.*

My enquiry led me to the conclusion that neither legal relations nor political forms could be comprehended whether by themselves or on the basis of a so called general development of the human mind, but that on the contrary they originate in the material conditions of life, the totality of which Hegel, following the example of English and French thinkers of the eighteenth century, embraces within the term 'civil society'; that the anatomy of this civil society, however, has to be sought in political economy. The general conclusion at which I arrived and which, once reached, became the guiding principle of my studies can be summarised as follows. In the social production of their existence, men inevitably enter into definite relations, which are independent of their will, namely relations of production appropriate to a given stage in the development of their material forces of production. The totality of these relations of production constitutes the economic structure of society, the real foundation, on which arise a legal and political superstructure and to which correspond definite forms of social consciousness. The mode of production of material life conditions the general process of social, political and intellectual life. It is not the consciousness of men that determines their existence, but their social existence that determines their consciousness. The changes in the economic foundation lead sooner or later to the transformation of the whole immense superstructure. In studying such transformations it is always necessary to distinguish between the material transformation and the economic conditions of production, which can be determined with the precision of natural science, and the legal, political, religious, artistic or philosophic in short, ideological forms in which men become conscious of this conflict and fight it out. Just as one does not judge an individual by what he thinks about himself, so one cannot judge such a period of transformation by its consciousness, but, on the contrary, this consciousness must be explained from the contradictions of material life, from the conflict existing between the social forces of production and the relations of production. No social order is ever destroyed before all the productive forces for which it is sufficient have been developed, and new superior relations of production never replace older ones before the material conditions for their existence have matured within the framework of the old society. Mankind thus inevitably sets itself only such tasks as it is able to solve, since closer examination will always show that the problem itself arises only when the material conditions for its solution are already present or at least in the course of formation.

The attentive reader will notice that science is here distinguished from ideology, but all versions of Marxist history of science generalise the position taken here about intellectual life and treat science as lying within the historical forces described by Marx in this passage, which is of fundamental importance to Marxist historiography.

Next to economism is the theory of mediation, according to which there are various degrees of relative autonomy, elasticity, lag and room for contradictions. There is ample warrant for this in the writings of Marx and Engels. For example, Engels wrote to Bloch in 1890:

> *According to the materialist conception of history, the* ultimately *determining element in history is the production and reproduction of real life. More than this neither Marx nor I have ever asserted. Hence if somebody twists this into saying that the economic element is the* only *determining one, he transforms that proposition into a meaningless, abstract, senseless phrase. The economic situation is the basis, but the various elements of the superstructure political forms of the class struggle and its results, to wit: constitutions established by the victorious class after a successful battle, etc., juridical forms, and even the reflexes of all these actual struggles in the brains of the participants, political, juristic, philosophical theories, religious views and their further development into systems of dogmas also exercise their influence upon the course of the historical struggles and in many cases preponderate in determining their* form. *There is an interaction of all these elements in which, amid all the endless host of accidents (that is, of things and events whose inner interconnection is so remote or so impossible of proof that we can regard it as non-existent, as negligible) the economic movement finally asserts itself as necessary. Otherwise the application of the theory to any period of history would be easier than the solution of a simple equation of the first degree.*

All of this falls within the general framework which asserts that:

> *The ideas of the ruling class are in every epoh the ruling ideas: i.e. the class which is the ruling* material *force of society, is at the same time its ruling* intellectual *force. The class which has the means of material production at its disposal, has control at the same time over the means*

> *of mental production, so that thereby, generally speaking, the ideas of those who lack the means of mental production are subject to it. The ruling ideas are nothing more than the ideal expressions of the dominant material relationships, the dominant material relationships grasped as ideas; hence of the relationships which make the one class the ruling one, therefore the ideas of its dominance.*

Marx stresses the historicity of all concepts throughout his writings, for example, in the *Grundrisse*:

> *This example of labour shows strikingly how even the most abstract categories, despite their validity precisely because of their abstractness for all epochs, are nevertheless in the specific character of this abstraction, themselves likewise a product of historic relations, and possess their full validity only for and within these relations. In the succession of the economic categories, as in any other historical, social science, it must not be forgotten that their subject here, modern bourgeois society is always what is given, in the head as well as in reality, and that these categories therefore express the forms of being, the characteristics of existence, and often only individual sides of this specific society, this subject, and that therefore this society by no means begins only at the point where one can speak of it* as such; *this holds* for science as well.

To summarise the less rigid of Marxist approaches to the history of science which fall within the domain of mediation theory, one should begin with the concept of labour and the labour process as the key to human history. History or historicity is, in turn, the key to everything: 'We know only a single science, the science of history'. The net effect of this approach is to broaden and deepen one's perspective: to root explanations in labour and the labour process, to treat concepts historically, to investigate connections and articulations as fully as possible and constantly to bear in mind that the arrow of causality moves from being to consiousness. This means that a number of distinctions on which the false self-consciousness of science depends are seen as permeable and interactive, for example, the distinctions between fact and value, substance and context, science and society, the context of origination and the context of justification. If one connects these perspectives to recent developments in the philosophy of science, a useful simplification would be to say that all facts are theory-laden, all theories are value-laden, and all values are derived from world-views or ideologies which permeate and constitute what count as facts, theories, priorities and acceptable scientific discoveries. A further consequence is that the sharp distinction between science and technology vanishes. All is mediation of social and economic forces involved in the production and reproduction of real life. Science is inside society, inside history.

At the extreme of the position I have described as the theory of mediation lies structuralist Marxism with its concept of immanent or structural causality in which the formal features of an intellectual sphere correspond to formal features of the base, but in the most arcane writings of the structuralist Marxists, the lonely moment of the last instance never comes. My own experience of the trajectories of structuralist Marxist writers is that they eventually find themselves moving into the New Right and are therefore not a reliable guide to this point of view.

At the outermost extreme of Marxist historiography lies the point of view of totality. The entire effort of Marxist writings in this tradition is to transcend the attempt to treat science in isolation from society. Science was seen as 'incapable of grasping reality as a totality'. At the extreme, the theory of totality argues that all aspects of reality are interconnected with and reflect all others. This is a point of view rather akin to Leibniz's monadology and it runs the risk of losing the directionality of causality from base to superstructure which is the bedrock or axiom of Marxist explanation. On the other hand, the point of view of totality insists on embedding ideas in society. As Lukacs wrote,

> *For the Marxist as an historical dialectician both* nature *and all the forms in which it is mastered in theory and practice are* social categories; *and to believe that one can detect anything supra-historical or supra-social in this context is to disqualify oneself as a Marxist.*

The appearance of Marxist history of science in the Anglo-American literature can be linked to a single catalytic event: the surprise appearance of a Soviet delegation at The Second International Congress of the History of Science and Technology in London in 1931. The delegation was headed by Nikolai Bukharin, Lenin's favourite, who was still years away from his own dramatic purge trial. Another contributor to the volume of their essays, *Science at the Cross Roads,* was N. I. Vavilov, an eminent plant breeder who died of persecution a decade later, while among the others, E. Colman survived and slipped into Finland decades later.

Far and away the single most important document in the Marxist historiography of science is the essay from that volume, 'The social and economic roots of Newton's "Principia"' by Boris Hessen. This is the *locus classicus* of the base-superstructure approach to the history of science, using the greatest work of modern science's most revered hero as its case study. Hessen argued that each of Newton's main theoretical preoccupations could be rooted directly and unambiguously in technical issues in his historical setting. He began by reviewing Marx's views from the Preface to *A Contribution to the Critique of Political Economy* (quoted above). The following sentence appears in bold: 'The method of production of material existence conditions the social, political and intellectual process of the life of society'. Moving on to Lenin, he claims that Marxism eliminates two main defects in previous historical writing:

> *Previous historical theories considered only the intellectual motives of the historical activity of people as such. Consequently they could not reveal the true roots of those motives, and consequently history was justified by the individual intellectual impulse of human beings. Thus the road was closed to any recognition of the objective laws of the historical process. "Opinion governed the world." The course of history depended on the talents and the personal impulses of man. Personality was the creator of history. The second defect which Marx's theory eliminates is that the subject of history is not the mass of the population, but the personalities of genius. The most obvious representative of this view is Carlyle, for whom history was the story of great men. The ideas of the ruling class in every historical period are the ruling ideas, and the ruling class distinguishes its ideas from all previous ideas by putting them forward as eternal truths. It wishes to reign eternally and bases the inviolability of its rule on the eternal quality of its ideas.*

He then traces the economics, physics and technology of the period of the English Civil War and the Commonwealth and dwells on communications, water transport, industry (especially mining) and war (including division of labour, army and armaments, and ballistics). He then concludes:

> *If we compare this basic series of themes with the physical problems which we found when analysing the technical demands of transport, means of communication, industry and war, it becomes quite clear that these problems of physics were fundamentally determined by these demands. We have compared the main technical and physical problems of the period with the scheme of investigations governing physics during the period we are investigating, and we come to the conclusion that the scheme of physics was mainly determined by the economic and technical tasks which the rising bourgeoisie raised to the forefront.*

We cite all these facts in opposition to the tradition which has been built up in literature, which represents Newton as an Olympian standing high above all the 'earthly' technical and economic interests of his time, and soaring only in the empyrean of abstract thought.

Then, casting his net wider, he allows for a less vulgar view than is often attributed to him:

> *It would, however, be too greatly simplifying and even vulgarizing our object if we began to quote* every problem *which has been studied by one physicist or another, and every economic and technical problem which he solved.*

He goes on to repeat that the economic factor is not the sole determining factor only the 'creation and recreation of actual life'. He then argues that it is important to 'analyse more fully Newton's epoch, the class struggles during the English Revolution, and the political, philosophic and religious theories that are reflected in the minds of the contemporaries of these struggles'. Hence, he discusses wider issues, including the history of concepts of energy, and the history of Luddism leading up to his own time.

I have considered this classical paper at some length, mostly because it is richer than is often supposed. Among the other writings in *Science at the Cross Roads*, Bukharin 's essay seems to me the most interesting. In discussing 'Theory and Practice from the Standpoint of Dialectical Materialism', he puts forward a theory of science which roots it in practice. He argues that the idea of the self-sufficient character of science 'science for science's sake' is naïve: it confuses the *subjective passions* of the professional scientist, working in a system of profound division of labour, in conditions of a disjointed society, in which individual social functions are crystallised in a diversity of types, psychologies, passions (as Schiller says: 'Science is a goddess, not a milch cow'), with the objective *social role* of this kind of activity, as an activity of vast *practical* importance. The fetishizing of science, as of other phenomena of social life, and the deification of the corresponding categories is a perverted ideological reflex of a society in which the division of labour has destroyed the visible connection between social function, separating them out in the consciousness of their agents as absolute and sovereign values.

He goes on to include scientific theories within the superstructure, and says that the 'mode of production' determines also the 'mode of conception'. Therefore, the 'highest forms of theoretical cognition' are included in historical analysis; '*scientific cognition* is the practice of material labour continued in particular forms (natural science)'.

In my view, this approach lays the foundations for a labour process perspective in the history of science in which the relations between theory and practice the connections or articulations of science are always to the fore as constitutive rather than contextual, while science itself is seen in terms of raw materials, means of production, purposive human activity and the goals or use values which emerge from the labour process.

The effect of the appearance of the Soviet delegation has been described as electrifying though not on the day, when hardly anyone responded. The papers were printed and made available in five days and appeared in book form ten days after they were delivered.

The historians of science whose work was most directly influenced in Britain were J. G. Crowther, a journalist and free-lance writer; Hyman Levy, a physicist; Joseph Needham, a chemical embryologist who became the historian of a massive work on *Science and Civilization in Ancient China;* and a polymath crystallographer, J. D. Bernal, who essayed broadly on the history of science, especially in his multi-volume *Science in History*. There were others, but I would say that the direct effect on historical writing (*pace* Needham) was not very great. It certainly did not influence the teaching of the history of science in the major British universities in the ensuing decades. Benjamin Farrington wrote interestingly on Francis Bacon, but the only noteworthy young historian of science in Great Britain, S. F. Mason, author of *Main Currents of Scientific Thought,* had to return to chemistry because he could not find work as a historian of science.

Instead, the history of science developed two deeply un-Marxist strands: the history of discovery and the history of ideas. Both were cut off from the study of the sorts of determinations which characterise the writings of Marx, Engels, Soviet writers and their English followers. There were a few English-speaking writers of note, especially Edgar Zilsel and Dirk Struik, but there was certainly no school or tradition. Indeed, an avowedly anti-Marxist historian of ideas, Rupert Hall, bragged in 1963 about how little Marxist writing there was and reiterated his position a decade later when he was badly out of touch.

Did it all drain into the sands? I would say no. Rather than look for directly Marxist historiography, it is worth noticing watered-down versions of it – linking ideas to their times but filtering out the nasty subversive and revolutionary potential of a fully Marxist analysis. I refer to the writings of Max Weber, Karl Mannheim and Robert K. Merton in sociology, the sociology of knowledge and the sociology of science. I will dwell on the last two, while noting in passing that Max Weber has been seen as the bourgeois Marx, placing ideas in history but rooting them in an irrational human nature rather than in the socio-economic base. Karl Mannheim, the father of the sociology of knowledge, has been called 'the bourgeois Lukacs', since his work was deeply influenced by the Hungarian Marxist dialectician whose writings on reification and the concept of totality were of fundamental importance in challenging economism or vulgar Marxism and providing a rich conception of the interrelations between forms of thought, ideas of nature and conceptions in science. Mannheim rooted ideas, including some ideas in the foothills of science, in human interest. He began with the concept of the value-ladenness of ideas which represented the interests of particular groups and then moved on to the sociology of knowledge. He wanted to determine the parameters of situationally conditioned knowledge versus detached knowledge. Mannheim acknowledged his debts to Marx and Lukacs, but his own work was distinct from theirs in that his goals were avowedly epistemological rather than political. His aim was to avoid taking sides and to find a standpoint from which to view knowledge as a social product, while the Marxist position is to assert that knowledge always takes sides and to support the relatively progressive tendency in any particular circumstance. Mannheim attempted to rise above the dangers of relativism into the observer's viewpoint of 'relationalism', while Marxists always see their work as fundamentally polemical and partisan.

Many adherents to the perspective of the sociology of knowledge have joined Mannheim in attempting to stand above the battle and have drawn on other social sciences, including anthropology, to enrich their views on scientific knowledge. The approach of the sociology of knowledge has grown dramatically in recent years as a way of rooting science in society without endangering the career or detached academic perspective of the investigator, drawing a firm line between theory and practice.

A similar path was taken by Robert K. Merton, the doyen of bourgeois sociology of science, whose original work in the 1930s was littered with footnotes and homages to Hessen. Merton focused on the origins, the class perspectives, the choice of topic, and other parameters of scientific knowledge while avoiding any commitment to seeing the resultant discoveries in ideological terms. The sociology of knowledge thereby became an elaborate study of the context of origination while carefully keeping away from the context of justification, the holy of holies which is so dear to non-Marxist philosophers of science. Within this framework of sociology of science as sociology of knowledge, quite subtle work has been done about scientific communities, patronage, honours, the culture of laboratories, scientific accountability (or the lack of it) to the rest of society, and other topics which take the existing mode of production as given.

There are traditions of Marxist history of science in Eastern Europe, but there has been very little communication between writers in the Soviet bloc and those in the West. There are also Marxist and communist writers in, for example, France, Italy and Germany, but Anglo-Saxon scholars have not been particularly powerfully influenced by them. The single exception to this judgement is the work of Michel Foucault, who wrote within the structuralist tradition and always claimed to do so in the light of Marxism. However, his voluminous and brilliant writings have always struck me as lacking historical specificity. He paints boldly and audaciously on large canvases, dipping into historical particularity at moments, but his concept of power is relatively a historical in that a particular claim about it could often be moved fifty years in either direction without affecting his argument. My own experience of most writing in the wake of Foucault's ideas is that the practitioners lack his brilliance, do not do their homework, and tend to drift into relatively apolitical stances, bedazzled by *belles lettres*.

The best defence of Foucault's writings is that, like a Marxist, he treats theoretical and practical knowledge as forms of power. His most illuminating work has been about human knowledge psychiatry, clinical medicine, penology and sexology. Yet he is un-Marxist in cutting these off from most of their articulations and in treating them formalistically. In this sense he is also anti-Marxist, or, at least, anti-reflectionist. Yet he leaves open the question of the appropriate conceptualisation of the relationship between knowledge and society: it remains to be investigated. For a Marxist, in the elusive last instance the question is not an open one. The arrow of causality must point, however circuitous its path, from the production and reproduction of real life to knowledge, whether theoretical or applied.

Would that I could cite a significant volume of avowedly Marxist writings in the English language to counterpose against the watered-down and disappointing versions discussed above. It could, of course, be argued that the flowering of Marxist writings in the later 1960s and early 1970s only glanced at problems which the history of the mode of production is not yet ready to solve. As Albert Camus, a fellow traveller of French Marxism, once wrote, 'One must imagine Sisyphus happy'.

In my own view, when future historians find themselves able to provide a richer Marxist history of science, the Marxist writer whose views they are likely to find most fruitful along with those of Marx, Lukacs and Marcuse is Antonio Gramsci. I shall cite two passages to give a flavour of his insight the first on objectivity and the second on science and nature: the idea of 'objective' in metaphysical materialism would appear to mean an objectivity that exists even apart from man; but when one affirms that a reality would exist even if man did not, one is either speaking metaphorically or one is falling into a form of mysticism. We know reality only in relation to man, and since man is historical becoming, knowledge and reality are also a becoming and so is objectivity, etc.

Matter as such therefore is not our subject but how it is socially and historically organised for production, and natural science should be seen correspondingly as essentially an historical category, a human relation. Might it not be said in a sense, and up to a certain point, that what nature provides the opportunity for are not discoveries and inventions of pre-existing forces of pre-existing qualities of matter but 'creations', which are closely linked to the interests of society and to the development and further necessities of development of the forces of production?

7

Rise of Paleontology

Introduction

Biologists were starting to realize, even before Darwin, that the crust of the earth was seething with the remains of plants and animals that had evolved, lived, and perished eons ago.

George Cuvier was among the important scientists working in this field. At the onset of his research, studies of prehistoric reptiles cultivated his appetite for paleontology. During the French Revolution, someone had sent a skull of what seemed to be a crocodile to the Jardin des Plantes. Explorers had found it many years earlier in the Netherlands on some land belonging to a "Dr. Goddin." Cuvier, in fact, proved that it was not a crocodile but an ancient lizard related to the modern monitor lizards. Cuvier incorporated many of his findings into his book *Research on Fossil Bones*. It was this book that, in effect, established paleontology as a separate science. Beyond this, Cuvier identified fossilized remnants from both oceanic reptiles and a potpourri of winged reptiles such as the pterodactyl, an extinct flying reptile. The keenness of Cuvier's "eye" was manifest when he identified a pterodactyl from only a drawing of the fossil remains.

In 1780, scholars identified the skull of a previously unknown dinosaur, which some amateur explorers had found in a quarry in the Netherlands. Eventually, science called this creature a "mosasaur." At this point Cuvier really began to exert his authority in paleontology. According to Cuviery 's measurements, the reptile stretched the measuring tape to nearly fifty feet. As is often the case in paleontology, the discovery was serendipitous. In fact, most scientists were unable to identify it, beyond realizing that it was an animal, until Cuvier realized, in 1795, that it was in fact a dinosaur. In the mid-nineteenth century, the paleontologist Richard Owen would propose the neologism "dinosaur" to describe the increasing number and diversity of such large reptile fossil discoveries paleontologists were unrelentingly uncovering. In 1799, a team of geologists on a routine expedition discovered an entire mammoth completely and faithfully preserved in the Siberian ice.

This was not to be the last of the colossal animals of prehistoric times that would come to light in this era. In at least one case, Cuvier, again, demonstrated that some earlier discovered fossilized bones, long thought by religious apologists to be the residue of an animal that had allegedly drowned in the great biblical Flood, were actually bones from an enormous salamander that had probably died from natural causes. Paleontology was ripening into a mature science. Well into the nineteenth century, the German naturalist Herman von Meyer found the first feather fossil. Scientists soon followed this bombshell with another, a nearly complete fossil of the same winged animal archaeopteryx. In his own speculations about evolution, von Meyer surmised that there were no literally "extinct" animals - simply animals that paleontologists imagined to be extinct but which were lurking elsewhere waiting to be discovered.

Type Theory

Still another of Cuvier 's enduring legacies was the "type" axiom. The type approach disallows comparisons between advanced life forms of independent species. For, according to Cuvier, all animals fall into one of four main "types" or groups: *Vertabrata, Mollusca, Articulata and Radiata.* Since the four groups are radically different in their overall design, one cannot understand an animal in one group by comparing it to an animal in another. Furthermore, Cuvier's various opinions on embryological development also contributed substantially to the contemporary dogma of evolution. Inevitably, Cuvier himself finally accepted the fact of species extinction and moved even closer to orthodox evolutionary teaching, although he never entirely relinquished his belief that the cosmos and species were evidence for the benevolent hand of the Almighty.

Rise of Catastrophism

Despite all of this, Cuvier certainly held a spiritual worldview that he sought mightily to reconcile with the scientific data. Today, most scholars agree that this reconciliation had taken place at least in Cuvier 's own mind by 1812, when Cuvier published his multivolume series, *The Fossil Bones of Quadrupeds*. In it he puts forth the "catastrophe" theory of evolution and presents guidelines for the study of comparative vertebrate paleontology. Cuvier became more and more consumed by fossil work as he developed his views on "catastrophism," the suggestion that periodically, the biosphere had eliminated entire species of plants and animals which the heavenly Father in his wisdom replaced in the ensuing years. This elegant philosophy, still prized even today in some quarters, urges that, periodically, some unprecedented catastrophe such as a major flood, an ice age, or the like annihilates every living species on earth. Professor Peter Ward of the University of Washington, for example, believed as late as the 1980s that ammonites, an ancient mollusc, had become extinct very gradually. Recently, however, even he has changed his mind, arguing that their extinction was in fact very sudden and probably due to a catastrophe of some kind. (Intriguingly, the story of Noah's ark was the prototype for all "catastrophe" ideas.)

Uniformitarianism

The scientists Hutton and Lyell, and Darwin himself, favored a competing hypothesis called "uniformitarianism." According to this idea, all of the observable features of the earth appeared bit by bit over vast stretches of time. Darwin incorporated a similar proposal into evolutionary teaching, arguing that the recasting of species also occurred gradually over inestimable numbers of years. Possibly because of Darwin's prestige, uniformitarianism would dominate geological and evolutionary thinking for years to come. Born in 1726, James Hutton, the child of a wealthy landowner, added still more fuel for Darwin, and science gained a priceless biological legacy in his geological commentary *A Theory of the Earth*. He argued that, because the evidence pointed so clearly to gradual revisions in living species, geology could offer no support whatsoever for the biblical doctrine of creation. In his view, the horizontal layer of fossils he found were the result of gradual deposits. The biblical Flood could not have produced them.

Exploration of the Seas

Soon scientists were starting to consider the possibility that biology had neither discovered all forms of life on the earth's surface, nor even begun to fathom the enigmas that lay beneath the seas. So it was that in 1815, the biologist Edward Forbes of Durham, England, demonstrated in a series of ocean voyages that living creatures thrived deep within the ocean. Scientists would do no further significant investigations in this arena until past the midpoint of the nineteenth century, when the zoologist Sir Charles Wyville of Scotland would start scavenger operations on the ocean floor that unearthed still more forms of life. And in 1872, the Scottish zoologist Charles Thomson was ready to begin a term as scientist-in-residence on board the Challenger, which was really the first significant oceanographic voyage in recent biology. Thomson spent four years on this trip gaining unique information not only on new life but about the depth and composition of the ocean itself.

Rise of Geology

The year 1799 was, defensibly, the most significant year for archaeology. In that year, soldiers in Napoleon's army uncovered a tablet that carried three varying forms of script. The Rosetta stone had returned to civilization. Soon, the verities scientists obtained from it would unlock the door to the translation of both Egyptian hieroglyphics and Sumerian cuneiform writing. One could scarcely consider it a surprise, accordingly, when, in 1802, the Austrian archaeologist Henry Rawlinson first translated a piece of cuneiform writing dating back to the ancient Persian empire, although no one got around to publishing his epochal translation until 1846.

William Smith

A commendable figure in connection with the field of geology, though he often receives scant credit, is civil engineer William Smith. Author of *The Geological Map of England*, Smith was born in 1769 to an impoverished family. Without obtaining any formal education, he became a surreyor, which spawned his lifelong search for biological knowledge in the strata of the earth. He was certainly the first scientist to use fossils for purposes other than for identifying organisms. He realized that fossils yielded another pattern of features altogether. If properly examined, it was possible to correlate individual fossils with certain "strata" or layers of the earth. By scrutinizing them and relying on the assumption that discrete strata correlated with specific periods in the earths history, geologists were able to date both the strata and the fossils themselves. Smith eventually published his views and theories in a huge atlas, and he is plainly the first scientist to appreciate the importance of fossils in dating distinct layers of the earth's crust.

Rise of Chemistry

Developments in other branches of science were occurring during this period as well. The physicists Henry Cavendish and Daniel Rutherford, for instance, both discovered nitrogen in 1772, while Scheele discovered the gas oxygen, which he dubbed "fire air," since fires thrive in oxygen. Although scientists had already begun to appreciate the connection to biology of these elements, both would prove to be of even broader significance for biology later on, when biochemists found that both nitrogen and oxygen were integrally related to protein structure. It should be noted that, although convention generally credits Joseph Priestley with the discovery of oxygen, recent historical reassessment has shown that Scheele probably did so concurrently, though he did not publish his inferences until after Priestley did.

Astronomy

In astronomy, Johan Bode of the Berlin Observatory devised a relation which stated that the distances of planets to the sun follow a predictable mathematical ratio, later called Bode's law. Bode also developed a theory about the make-up of the sun similar to William Herschel's. Additionally, Pierre Simon Laplace of the French Academy of Sciences, famous for his work in probabilities, proved that even the entry of another planet into the solar system would produce only an infinitesimal deviation in the orbit of any other planet. In the ensuing years, a decade after the French Revolution, Laplace would collect, in several volumes, all of the advances in astronomy to date. They would appear in his magnificent *Mehanique Celeste* ("Celestial Mechanics").

Physics

Henry Cavendish also proposed that electricity was in reality a motion of a "fluid," a dogma that the American physicist Henry Rowland discredited in the nineteenth century. Later, Cavendish would measure the gravitational constant "g." The idea involves suspending two smaller lead weights next to two larger lead weights. The gravitational force of the larger attracts the smaller weight, thus twisting the suspending wire. The force required to twist the wire yields the gravitational constant. The ratio of the large to small objects yielded g. Consequently, Cavendish concluded the research in physics that Newton had begun. Using the gravitational constant, he revealed that the earth is about six times denser than water. As in prior epochs, the imperial progress of physics only inspired biologists to continue to try and apply the precepts of physics to biological phenomena.

Medicine

In medicine it was the age of Mesmer, the German physician who first used hypnotism then called "animal magnetism" or "mesmerism" to treat illness, an approach which Sigmund Freud took up years after, in the nineteenth century. By the mid-nineteenth century, the Scottish physician Sir James Braid had renamed the effect "hypnotism," and even tendered a theoretical mechanism explaining how it functioned. Hypnotism would rapidly find a secure niche in the physician's repertoire. Roughly coincidentally, Sir Percival Pott of St. Bartholmew's Hospital anticipated the modern science of environmental. medicine when he found an unusually high incidence of scrotal cancer among British chimney sweeps, guessing correctly that soot was the culprit.

Age of Kant

For philosophy, this period was emphatically the most critical in history. It was in 1781 that the German philosopher Immanuel Kant published his august *Critique of Pure Reason*, which triumphantly fused the empirical tradition of Locke, Berkeley, and Hume with the rationalist tradition of Descartes, Leibnitz, and Spinoza.

Botany and the Rise of Population Theory

In the ensuing ears, the Reverend Gilbert White added to the development of botany with his *Natural History and Antiquities of Selborne* a Superbly organized group of descriptions of flora and fauna of the English town of Selborne. Published in 1789, this volume helped establish the importance of environmental and ecological issues in biology. Soon would follow the magnificent reflections of Thomas Malthus in 1798. Malthus was born in 1766, the offspring of a well-to-do landowner in England. He prepared himself for the priesthood at Cambridge, although his professional passion was economics. His father had worked with the French Enlightenment philosopher Jean-Jacques Rousseau and had taught his son the value and justice of redistributing wealth for the net benefit of the masses. Thomas Malthus *Essay on the Principle of Population as It Affects the Future Improvement of Society* appeared under a pseudonym in England, as Malthus was troubled by the possible repercussions of such a drastic social tenet. His thesis in this book was that the principal cause of misery in the world was the careless and wasteful use of the planet's resources. He proposed that only famine or war could control world population a population that was rapidly spiraling out of control.

From the vantage point of biology, the salient issue in this book was Malthus's notion that competition in society would inevitably and methodically weed out the substandard and less "fit" classes through crime and evildoings a recommendation that Darwin would exhaustively analyze.

Entomological Taxonomy

No one had done anything of striking merit in entomology since the epochal findings of Rene de Reaumur in 1734. The zoologist Johann Fabricius (a different scientist from his namesake of nearly two centuries past) ended the drought in 1775, when he published his *Systema Entomologiae,* or "Systematic Entomology." Born in Denmark in 1745, Fabricius was a physician's son. He enrolled at the University of Copenhagen in 1763. Later he went to the university at Uppsala, where he befriended Linnaeus, creating a relationship which would last for many years. Sometime afterwards, the University of Kiel appointed him to a professorship. Even with the low salary they offered, Fabricius somehow managed to spend most of his semesters reflecting and visiting his friends in Paris.

With his *Systema Entomologiae,* Fabricius had taken an impressive step forward in two branches of biology insect investigations and taxonomy. In large part he applied Linnaeus's methods to entomology. Even the titles of his publications were similar to those of Linnaeus. In this dissertation he was able to productively and usefully classify species and varieties of insects using variations of mouth structure rather than wings as his basis.

Sprengel and Insect Reproduction

Without letting the momentum ebb, the German biologist Christian. Sprengel, continuing the work of Kolreuter, explained still more of the ways in which both insects and wind could, by carrying pollen, fertilize plants. Christian Conrad Sprengel was born in 1750 at Brandenburg to a pious and devout family, his father having served as a lay vicar for many years. His early education was in language and theology as well as philosophy. For many years thereafter he served as a headmaster of a Berlin school. However, his disposition was mercurial and not easily tolerated. His personality conflicted with both parents and other faculty at the school. He was also, so far as anyone can tell, a perfectly awful teacher. Inevitably, the school had to ease him out, albeit with a full pension, in 1794. He lived most of his years alone, passing away in Berlin in 1816. His studies, conducted mostly at home, consisted of assembling volumes of information on florescence in plants as well as a number of anatomical descriptions of a multiplicity of flora.

Most of his ideas on plant fertilization, particularly the above-cited conjecture that insects and wind could carry pollen, were more or less in line with Kolreuter's and still hold true today. As Sprengel himself describes things in his book *The Secret of Nature Discovered in the Structure and Fertilization of Flowers*.

In the summer of 1789 I examined several species of iris, and found that these flowers could be fertilized in no other way than by insects. Then I searched for other flowers whose structure should show the same thing. My studies convinced me that many, perhaps all of the flowers which secrete nectar are fertilized by insects.

As was true of so many scientists of this period, Sprengel saw the created order as one grand proof of the reality and munificence of the Lord. Like Kolreuter and Mendel, Sprengel, too, had to stomach neglect at the hands of his fellow scientists. In part this may be due to the rather obvious theological messages in his writings, which served only to alienate thinkers who, in this period, were desperately trying to comprehend the heavens "mechanistically." The supposition that some deity had designed all of nature to reach some final state was too reminiscent of Aristotle and medieval theology. Fortunately, Darwin himself rescued Sprengel from oblivion. He based many of his own conclusions about evolution and natural selection on Sprengel's investigations of the anatomy of flowers and plants as well Sprengel's observations on the relationships between insects and flora.

Conclusion

Above all else this was the era of Cuvier, which saw the rise of paleontology and the discovery of innumerable important animal remains such as the mosasaur. This era also witnessed the gradually escalating competition between the "catastrophist" and "uniformitarian" ideologies in evolutionary theory a controversy that has not resolved itself even today. Not surprisingly, the "cousin" science to paleontology, geology, began to evolve with the work of Jean Guettard, who first demonstrated the existence of an ancient ice age.

Other sciences too were advancing rapidly. The discovery of Bode's law in astronomy, the work of Cavendish in physics, and the work of Scheele in chemistry all bear witness to that. In medicine it was the era of Mesmer, and the German virtuoso Immanuel Kant dominated philosophy. Botany too made progress in this period with the work of Gilbert White, while Thomas Malthus would begin his epochal reflections on population growth. Invertebrate zoology, particularly entomology, pushed ahead with the efforts of Sprengel and Fabricius.

8

Darwin and His Experiments

Introduction

A graduate of Andover Academy, became professor of vertebrate paleontology at Yale, as the number of professional, academic positions for scientists gradually, if sluggishly, continued to expand. He was one of the very first to be appointed to such a post, as well as to the U.S. Geological Surve. He dedicated the balance of his days to searching for dinosaur fossils, in the, American West and noting their implications for the ancestry of all species, including man. He described many of his ideas in his book, *Dinosaurs of North America*.

During the initial decades of the Victorian era, scientific institutions were proliferating and thriving. The Royal Society's membership list registered strong gains throughout the nineteenth century, as did that quintessentially Ametican scientific institution, the Smithsonian. In 1823, for instance, the American zoologist Spencer Baird, founder of the Woods Hole Laboratories, while head of the Smithsonian Institution, systematically compiled and preserved huge numbers of animals in the Smithsonian collection, a collection that is recognized today worldwide. Baird is also the author of the *Catalogue of North American Birds*.

Yet while the social structure of science was growing more organized and professional, other forces acted to retard the growth of science itself. Undeniably, theology had not yet lost its grip on science. In his inaugural lecture at Oxford University in 1820, for instance, the eccentric naturalist William Buckland strongly insisted that the goal of geology is and ought to be nothing else but to confirm biblical claims an event suggestive of the controversies that arose during the twentieth century over evolution versus "creation science." In 1823, he published his *Observations on the Organic Remains Contained in Caves, Fissures, and Diluvial Gravel and on Other Geological Phenomena*. Predictably, this tract "confirms" Buckland's preconceptions. By systematically twisting evidence to his favor when it could be twisted and ignoring evidence that could not be twisted to bolster his argument, he "proved" that the biblical Flood had happened, using water-level marks in caves to prove it. He also "proved" that it had occurred about 6,000 years ago. When Darwinism first appeared, and for many decades after, religious fundamentalists would try to force it out of the curriculum of the public schools because it appeared to contradict Genesis.

Mechanism

This era witnessed an ever-escalating controversy over mechanism, the idea that all biological phenomena can be explained purely through the laws of physics. Scientists like Claude Bernard of the Sorbonne, in his *Introduction to ihe Study of Experimental Medicine*, would argue that the philosophy of mechanism is true and that the concept of a "vital force" is clearly so much useless, abstruse baggage. Science plainly did not need it.

Darwin himself made his own contribution to this controversy. In 1872, his book *The Expressions of the Emotions in Man and Animals* appeared. In it Darwin added his name to the list of defenders of the "mechanistic" analyses of the evolution of the mind. Here he defends the notion that human emotions, for instance, descended directly from analogous behaviors in more primitive species. And, in his studies of the circulation of the blood, Karl Witzenhausen of Germany argued that bodily activity was explicable on a purely mechanical basis.

Philosophically, the German theorist Ernst Haeckel, too, was a materialist in the spirit of the pre-Socratic philosophers. There was no deity, and the universe was fully explicable with the laws of physics. Life too began through blind chance, probably from some random combination, of chemical elements. Haeckel further angered traditional fundamentalists when, like Darwin, he proclaimed that there was no vital difference between the mind of man and the minds of lower animals, simply a distinction in degree. In fact, psychology was not and could not be anything more than a branch of physiology.

As things stand today, the controversy over mechanism has hardly abated. Many important philosophers, such as the British analytic philosopher Elizabeth Anscombe and D. M. Armstrong of the University of Adelaide, still argue that the old Cartesian idea of the "soul" has been completely refuted. Yet many philosophers believe today that this is not necessarily true. This group of thinkers, many heavily influenced by the legendary philosopher Ludwig Wittgenstein, have offered enormously powerful arguments against mechanism in any form.

Paleontology and Richard Owen

In 1822 in Tilgate Forest, the amateur Zoologist Mary Mantell, a confidante of Charles Lyell, found the fossil of an ancient lizard, the later-named *Igunadon*. This was a pivotal point in the development of paleontology. For although dinosaur remains had appeared earlier, no one had recognized them as such. In fact, the term "dinosaur" did not even exist at this point. It was Richard Owen who coined that term to apply to the giant reptiles that had ruled the earth many millions of years ago.

Richard Owen was born in 1804 in Lancaster, England, the offspring of a local retailer. Distinguished as he became in succeeding years, he gave no signs of special talent in early life. Expecting to go no further in his vocation than the local pharmacy, he began an apprenticeship in that field, yet a passion for science and medicine drove him constantly. Finally, he headed for the University at Edinburgh to learn medicine, spending his spare hours on anatomy.

As an anatomist he was without peer in his native land. His anatomical descriptions were scrupulously accurate, as were the drawings that accompanied and recorded them. He studied biology in Paris under Cuvier and became director of the natural history section of the British Museum in London, remaining in that position until past eighty years of age.

Unlike so many of his predecessors, he was not only an illustrious experimenter but a theoretician as well, and he indulged in a wide array of speculative explanations of what he had seen in his investigations. Following Cuvier, he scanned the same organ throughout the animal kingdom to try and understand how the organ had recast itself in the course of evolution. In this manner, he established that despite remarkable changes that evolutionary processes had caused, the same organ could do the same thing or perform the same function in animals separated by immense evolutionary gaps. Both the flying fish and birds used their limbs as wings, for instance, and the lungs of mammals and swim bladder of some fish also performed similar functions.

Owen on Homology and Analogy

In 1858, Owen published *On the Classification and Geographical Distribution of the Mammalia* and, two years after this, *Systematic Summary of Extinct Animals and Their Geological Relations*. He also made a valuable contribution when he published *On the Archetype and Homologies of the Vertebrate Skeleton*. Here he envisioned the skull as emanating from a series of adaptive over haulings of the vertebral column. This is where he made his momentous distinction between homologies and analogies. In his view, and in the view of all subsequent biologists, two body parts are "homologous" when they have the same evolutionary heritage even if their functions have little to do with one another. On the other hand, they are simply "analogous" if they have similar functions but divided evolutionary histories. The penis and the clitoris are homologous, for instance. This view was strikingly similar to that of Goethe's at the dawn of the nineteenth century.

These notions still play an integral role in modern comparative anatomy, and any sophomore year biology major will very quickly gain fluency with them. Beyond this, Owen 's anatomical concepts of homology and analogy appear, quite soundly, in most of the initial arguments in favor of evolution. It is also arresting that Owen had originally been in favor of Darwin's views. What he could not allow, in spite of this, was the further contention that man was part of the evolutionary pathways. Thus, he differed more with Darwin's book *The Descent of Man* than with *The Origin of Species*. Inevitably, this preoccupation with the distinctive status of man, traceable, no doubt, to Owen's religious beliefs, led him to break completely with Darwin.

Agassiz's Contributions to Paleontology

Paleontology pushed further ahead when the Harvard zoologist and geologist Jean-Louis Rodolphe Agassiz, a student of Cuvier 's, published the book *Recherches sur les Poissons Fossiles*, or "Studies of Fossil Fishes," again unveiling the strength of geology to illuminate biology. In fact, Agassiz compiled a massive work on glacial geology, titled "Studies on Glaciers," in which he waxes eloquently on the formation and movement of glaciers, revealing as well his bias in favor of a biblical view of creation:

> *gigantic carnivora, was suddenly buried under a vast mantle of ice, covering alike plains, lakes, seas, and plateaus. Upon the life and movement of a powerful creation fell the silence of death.*

Geological Work of Lyell

Closely related to paleontology is geology and the work of Charles Lyell. The son of a Scottish farmer, Charles Lyell entered the world in 1797. Because Lyell's father was also a student of biology, he provided an atmosphere conducive to developing his son's boundless native talent. Lyell graduated from Oxford University and practiced law for a while, until geology posed too strong an enticement to him.

Of particular note is the fact that he went on many voyages, surveying the fossilized forms of life that dotted the earth's crust. That, in turn, provided the theoretical. basis for Darwin's revolutionary conviction that higher forms of life descended from lower ones.

Lyell studied the potent forces that had served to shape the nature of the earth over eons. He also concluded that the same forces were operating at the same rate during his time as they had millions of years ago. That fact, he argued, precluded any reason to accept the biblical tenet of "catastrophism," or the conviction that the earth and the many forms of life on it had altered drastically and abruptly at various stages in the earth's history. For he insisted, in fact, that the fossil record showed that the reverse was the case the history of species was a study in gradual change. Many of these suggestions appear in his *Principles of Geology* (1830-1833). In these volumes, he also disavows the Lamarckian chimera of the inheritance of acquired characteristics.

Several years later, Lyell added substantially to the biological and geological sciences when he found indications of the existence of an extraordinarily early form of human being in the unearthing of some, stone tools.

Not surprisingly, such an astute scientist as Lyell had powerful supporters. The second part of Darwin's geological studies, for example, was even more significant than the first in that Darwin offered evidence confirming the views of Lyell about the existence of the Pleistocene, a subdivision within the Cenozoic period, the most recent era of geologic time. The Pleistocene is also the era where most modern flora and fauna appeared. Darwin also confirmed Lyell's views about other periods in the earth's history. This reassessment also tended to confirm the existence of an ice age in the earth's distant past.

Chambers as an Influence on Darwin

Another classical predecessor of Darwin was amateur geologist Robert Chambers (author and publisher of *Tracings of the North of Europe*) with his book *Vestiges of the Natural History of Creation*. Although there were several inaccuracies in this book, the deliberations in it were sound enough. His support gave encouragement to both Darwin and Wallace to push on with their development of the theory of evolution.

Progress in Botany

Botany was adding to its store of data as well. In 1823, the German botanist Nathanael Pringsheim became, arguably, the first biologist to make the mastery of algae his exclusive preserve. With his research, science for the first time came to know something about the biochemistry as well as the taxonomy of algae. In his classic study of the algae *Vaucheria*, he clearly set out how fertilization occurs. In addition, a quiet New Englander named Asa Gray was completing his own analysis of plant structure and function. Soon he would publish the *Elements of Botany*, which ultimately established him as one of the preeminent botanists of the past. He followed this with his *Manual of Botany*, which included a catalogue of all of the known plants in the northern United States. Beyond this, Gray was a well-known supporter of Darwin, opposing even Gray's own colleague at Harvard, Louis Agassiz. After many years of tireless exertion, he collected all of the evidence that he believed supported Darwin in his book *Darwiniana*. Notably, Gray avowed that there was no real conflict between evolution and theology.

Botanist Discovers Brownian Motion

The early part of this era was one of the exceedingly few times in history when a scientist made a major discovery in a field other than his own. In 1827, the Scottish botanist Robert Brown, while doing some routine microscopic analysis, noticed that particular liquids contained vanishingly small particles that remained in constant motion without any outside force, such as stirring, being applied to them. Science in succeeding decades dubbed this movement Brownian motion, which is how science still refers to it today in Brown's honor. During the early decades of the twentieth century, scientists and physicists would offer this phenomenon as proof of the reality of molecules.

Brown was born in 1773, the son of a Scottish clergyman. He began his training in medicine at the University of Edinburgh and served as a military physician. This phase of his career over with, he devoted himself to botany and did a considerable amount of investigation in Australia.

Brownian motion was not Brown's only contribution; among other accomplishments he described the cytoplasm of cells quite precisely in his monograph *Microscopic Observations on the Pollen of Plants*. He is also well known for having discovered the cell nucleus. Another of his celebrated contributions consisted of his meticulous accuracy in cataloguing the various families of plants. An excellent example of this is his exploits with the Asclepiadaceae the milkweed family. There are a bewildering number of plants in this family, found mainly in Africa, and some of them are cactuslike. Brown was also what one might call a "geographer" of plants, having studied the way plants are scattered under assorted climactic conditions.

But Brownian motion endures as his most heralded contribution. Oddly, the actual size of his published conclusions is modest, though some of his papers were significant enough to be published in Germany—then the world leader in science. Nonetheless, the scientific community fittingly regards him as one of the notable botanists of history.

Darwin and Plant Evolution

Many tend to overlook the fact that Darwin concerned himself not only with animal but also with plant evolution. Indeed, soon after he published *The Origin of Species* he insisted that specialized parts of flowers must have evolved to allow the most efficient pollination by insects. Eventually, he followed this argument with *The Different Forms of Flowers on Plants of the Same Species*, in which he elaborated on the idea that one could resolve the radically distinctive variations in flowers by suggesting that plants had divergent evolutionary "purposes."

Darwin and Rise of Evolution

One of the first major steps toward the legendary theory of evolution took place when the entomologist Henry Bates engineered his epochal studies of insects in South Africa. His finest work here is his study of the *Heliconidae,* a type of butterfly. Though today's generation of biologists perhaps does not give him the appreciation he merits, they concede that he contributed to the scientific world's eventual endorsement of Darwin's tenets.

Darwin was born in 1809 at Shrewsbury in western England, the heir of the physician Robert Waring Darwin. Charles Darwin was one of eight children, and his education was traditional. He studied the classics and later went to the University of Edinburgh to begin his schooling in medicine. But he shortly left to train for the ministry at Cambridge. Natural history soon absorbed most of his attention At Cambridge University he studied theology, but during his spare moments he collected insects and read virtually everything he could find on geology. He had the happy fortune, too, to apprentice at Cambridge with Adam Sedgwick, a distinguished teacher of geology during his time.

Doubtless, Darwin's first important move toward formulating the theory of evolution, influenced somewhat by the geological findings of the Victorian geologist Charles Lyell, was Darwin's decision to travel on the *H.M.S. Beagle*. Starting in 1831, the journey was destined to last five years. England herself had commissioned the *Beagle* to sail around the tip of South America and the globe. The British Crown had charged its staff with drawing maps, discovering untouched locales, measuring distances, and so forth. Recognizing Darwin's eminence in science, the ship's officers immediately appointed him "scientific officer," though the position carried no salary. By 1835, they had arrived at the Galapagos Islands, where Darwin noted that certain finches appeared to have evolved from some common ancestor other than what then existed in South America.

In 1839, Darwin married his cousin Hannah Wedgwood. Her vast fortune, spawned by her family's thriving ceramics industry, allowed Darwin to spend the remainder of his life in security, able to pursue his interests full time, despite the fact that he suffered from ill health off and on for much of his life.

In the ensuing years he published his epochal *Journal of Researches into the Natural History and Geology of the Countries Visited during the Voyage of H.M.S. Beagle*. This book describes in painstaking detail Darwin's voyages over the five-year period, including his surveys of an enormous variety of animals and plants as well as numerous fossils. Recognizing the significance of geology for biology, Darwin had made it a point to thoroughly explore the geologic features of all of the areas he had visited. He reported these features in his book as well. When the voyage ended, Darwin read Malthus 's august writings on population and pondered his thesis that innumerably more individuals are born than can survive. Accordingly, the weaker ones had to perish in the competition for food. In 1840, Darwin published *Zoology of the Voyage of the Beagle*, in which he outlined in vivid detail all of the fauna and flora he had assembled on his travels.

By 1842, Darwin was confident and ready to publish his book *The Structure and Distribution of Coral Reefs, Being the First Part of the Geology of the Voyage of the Beagle*. Here he added, quite inadvertently, to the science of taxonomy, when he segregated coral reefs into three categories. There was, however, beyond the above inquiry, comparatively little advance in the science of classification that came from Darwin's pen. In this treatise, he also stated his own beliefs about the formation of atolls and coral reefs, via subsidence of islands in those locations. He theorized that atolls or coral reefs had formed when the land on which they were built gradually sank deep into the sea.

Origin of Species Appears

The consummate scientific event of the nineteenth century occurred the publication of *On the Origin of Species by Means of Natural Selectilon, or the Preservation of Favoured Races in the Struggle for Life.* Here Darwin explains in painstaking detail the premise of natural selection, mentioned earlier. He argued that nature, rather than a deity, selects among the offspring those best fit to survive. He noted, among other things, that the parents of most species produce more offspring than can possibly survive. That, in turn, leads to competition for shelter, food, water, and so forth. Then a random factor surfaces. Given that offspring vary in their properties quite haphazardly, those with the qualities exceptionally suited to the environment will survive. Some might be stronger, or have better hearing, or a body color which better blends in with the environment to hide them from predators. As a consequence, plants and animals with the more suitable qualities will transmit them to their offspring. In his second major tract, *The Descent of Man*, published three years after *The Origin of Species*, Darwin argued that man first descended from an apelike ancestor. He also shows how the operation of natural selection affected the actual development of various known species.

A surprisingly cosmopolitan aspect of this book is that Darwin announced that the "mental life" of *Homo sapiens* is not as unique as many theologians had always presumed. He argues legitimately that such psychological properties as memory, imagination, curiosity, and so forth exist in varying degrees in lower animals as well. Recent experiments with chimpanzees and language-learning inform us that Darwin's thinking here was both lucid and profound.

Sexual Selection

The Origin of Species did not end Darwin's labors. A few years later he published *The Variation of Animals and Plants under Domestication*, which was a further elaboration and discussion of chapter one of *The Origin of Species*. *The Descent of Man and Selection in Relation to Sex* shortly complemented *The Origin of Species*. In this treatise, he focused on the evolution of human beings from lower life-forms. He also demonstrated the truth of sexual selection. He devised this notion because he believed that ordinary selection processes could not elucidate secondary sexual qualities, for example, brilliant colors in an assortment of animals including butterflies and birds, antlers in the stag, and the like. Instead, a unique kind of competition between males emerged to win favor with females. In this way, the strongest and most handsome males were able to propagate and transmit their characteristics to the next generation. As Darwin described the phenomenon.

This form of selection depends, not on a struggle for existence in relation to other organic beings or to external conditions, but on a struggle between the individuals of one sex, generally the males, for the possession of the other sex. The result is not death to the unsuccessful competitor, but few or no offspring. Sexual selection is, therefore, less rigorous than natural selection.

Later, in his declining years, Darwin also began his *Animals and Plants under Domestication*, a two-part text that included a detailed description of his biological investigations on both plants and domestic animals. Beyond any doubt, the most important aspect of this effort was that it appended some valuable ideas to the suggestion that higher forms descended from lower ones, even hinting at a rudimentary notion of heredity. In it, he implied that parents could transmit qualities directly to their progeny under the control of a conglomeration of environmental factors, including infrequent use of a limb, climate, the food supply and so forth—a remarkably Lamarckian strand in Darwin's thinking.

Storm of Controversy

The immediate aftermath of Darwin's ideas was not entirely favorable. It scarcely needs saying that the religious ramifications were and still are disturbing to many. How could one now reconcile Darwin with Genesis? Yet Darwin was not the only source of trouble for believers. The German theologian David Strauss and the French thinker Ernest Renan had both disclaimed the historical accuracy of the Bible, exposing many of its inconsistencies. Without reservation, nevertheless, the most troubling aspect of Darwinism for the orthodox was the direct insinuation that no deity governed the cosmos. Instead, the universe resulted from blind, random chance. There were no assurances, no fixed celestial truths not even fixed moral truths.

The social implications of Darwinism went even deeper. Ultimately, a doctrine would appear called "social Darwinism" which applied Darwinian principles to the evolution of societies and moral codes. Those moral codes were "true" which assisted a human being or society to survive; they were not true because a deity had decreed them to be true from on high. Among other things, the Social Darwinists swore that the white race had proved its superiority to the other races by the "success" of Western Civilization. Among the exponents of this teaching, though he never espoused the most racial applications of the view, was Herbert Spencer. Like Huxley and Haeckel, Darwin transformed Spencer's thinking, and he universalized the concept of evolution so that, in his view, everything in the universe was the product of evolution.

Huxley Supports Darwin

The tempest swirling around Darwinism soon profoundly touched men like T. H. Huxley, who, along with Asa Gray, became one of Darwin's most conspicuous defenders. Huxley was of inestimable help to Darwin and added some much-needed rationality to the whole question of evolution versus special creation when he began his crusade in favor of Darwinism. Darwin himself had laid down the gauntlet in *The Origin of Species.*

I can entertain no doubt, after the most deliberate study and dispassionate judgment of which I am capable, that the view which most naturalists until recently entertained, and which I formerly entertained namely, that each species has been independently created is erroneous.

Huxley entered the world in 1825 and, in childhood, displayed stupendous scientific as well as philosophical talent, devouring every book he could locate. He began systematic biological study at Sydenham College, shortly receiving his bachelor's degree in 1845 from the University of London. Two years after that he completed his medical degree.

He immediately joined the British Navy as a medical officer stationed on the *Victory*, transferring after less than a year to the *Rattlesnake* because of hostilities on the former ship as well as because the latter ship was heading to Australia. That nation would provide him with a chance to inspect a range of flora and fauna in its rich biological environment. He accumulated enormous amounts of biological data on this trip, and in 1854 he published *Oceanic Hydrozoa* under the auspices of the famed Royal Society of London.

On the one hand, Huxley at first did not accept evolution. On the other hand, he did not think that Lamarckianism was a practical alternative either. Nor could he accept the biblical view. This philosophical stalemate ended, nonetheless, when he met the great philosopher Herbert Spencer in 1852. Since Spencer was a staunch advocate of evolution, Huxley's many discussions with Spencer shortly turned him toward Darwinism. In the end, he became such an advocate of evolution that he essentially became Darwin's publicist. When *The Origin of Species* appeared in 1859, Huxley's name would be everlastingly linked with Darwin's.

The controversy over Darwinism had reached its zenith. So high did passions run that in 1860 Huxley had it out with Bishop Wilberforce, then Bishop of Oxford and a professor of mathematics, at an impassioned meeting of the British Association for the Advancement of Science. The Bishop scored first when he asked Huxley whether such evolution was through their grandfathers or grandmothers. Although he forswore an actual proclamation of atheism, Huxley replied that tie would rather have an ape as an ancestor than the bishop, for such a beast "would not misapply his intelligence to joke about such a serious matter." That reply drew loud and long applause from the spectators, hinting, perhaps, that the grip of biblical Fundamentalists was beginning to weaken.

Beyond this, Huxley did say that "there is no evidence of the existence of such a being as the God of the theologians." He then declared that Christianity was a compound of some of the choicest and some of the worst elements of Paganism and Judaism, molded in practice by the innate character of specific people of the Western World.

In fact, it was Huxley who coined the term "agnosticism." With it he declared his hatred for the blind, dogmatic certitude he saw around him. Because of his uncompromising defense of Darwin, he earned the nickname "Darwin's Bulldog." Still, Huxley did much elegant research of his own in biology over a span of many years, chiefly in comparative anatomy. But his vigorous and long defense of Darwin is one of his chief legacies to biology.

The bishop was not the only one to vent his feelings about the villainous teachings. On the other side, some scientists sided with Wilberforce. Among others, the American biologist Louis Agassm attacked the theory, asserting that it was absurd and impossible. Instead, he proposed reinstating the venerable theological notion that God created all of the species separately (though his own findings actually substantiate evolution).

Wallace and Evolution

Alfred Russell Wallace undeniably deserves mention as the cocreator of the theory of evolution. Born in 1823, he lived a long and full life, dying in 1913 after spending his years in the service of science. Besides evolution, he devoted a considerable amount of attention to the distribution of animal species all over the globe. At the age of fourteen, he and his brother William began to master surveying, a prevalent occupation at the time. However, his surveying projects soon led him far afield, into both astronomy and botany. By the mid-1800s he had become a disciple of the brilliant Thomas Malthus, coming, like Darwin, under the influence of Malthus's essay *On Population*, thereby further cementing his obsession with biological questions. In April of 1848, his thirst for biological knowledge was nearly insatiable; he left England with his younger brother Herbert to explore the flora and fauna of South America. Sadly, disaster struck when his ship caught fire on the way back to England. Although he survived, he lost virtually all of the specimens he had collected as well as his own diaries. Undaunted, he relied as best he could on memory and authored both *Travels on the Amazon and Palm Trees of the Amazon*.

At forty-six, he married and began further biological quests in London, where he gave final form to his views on evolution. He wrote these up in his book *Contributions to the Theory of Natural Selection*, published in 1870. Possibly his most renowned volume, the above book notwithstanding, was his *Geographical Distribution of Animals*, which he finished in 1876. In this book he divided the planet into six land masses, all having their own characteristic life-forms. The "Paleartic," which encompassed a substantial part of

Asia and almost all of Europe, was home to reindeer, pigs, hawks, pigeons, dogs, rats, cows, cats, other varieties of deer, oxen, bears, and other animals. The "Neartic" included North America and housed large animals like foxes, bears, and elks. He differentiated "Ethiopia," the rustic habitat of gorillas, lions, tigers, giraffes, and rhinoceri. The Oriental region included China and Southeast Asia and its natural inhabitants elephants, flying foxes, and orangutans. In Australia and New Zealand he surmised there dwelled abundant marsupials. Lastly, in the "Neotropical" zone, consisting primarily of South America, dwelled sloths, tapirs, monkeys, bats, and other animals. Of course, he thoroughly scanned the accomplishments of his compatriot Darwin, even writing a book about him called *Darwinism*.

Wallace would achieve a level of recognition equal to Darwin's when, in 1858, the secretary of the Linnean Society read from several letters traded between Alfred Wallace and Darwin, as well as from an unpublished paper of Darwin's. In all of these manuscripts the authors described the thesis of evolution in considerable detail, even devoting some considerable space to the concept of natural selection. These letters and other evidence made it clear that Wallace had come up with the main ideas of evolution independently of Darwin.

Taxonomy: Invertebrate

Debates over evolution did not comprise the whole of nineteenth century biology even if it sometimes seems that way. Since Linnaeus's day, scientists had been gradually building on the foundations of taxonomy set down by Linnaeus. One significant extension of that science occurred in 1872, when the Breslau botanist Ferdinand Cohn, a pupil of Johannes Muller, published the first of a three volume series devoted to arranging bacteria into genera and species titled *Bacteria, The Smallest of Living Organisms*. This volume summarized all of Cohn's epochal research in the embryonic field of bacteriology to date. The classification of bacteria continued with the deliberations of the Danish bacteriologist Hans Christian Gram, who, in 1884, invented a stain that he used to split bacteria into a gram positive group and a gram-negative group as he called them. The former group absorbed the stain, and the latter group did not. The immediate practical significance of this admittedly odd principle of grouping was not clear. By the 1940s, however, bacteriologists found that the classification was indeed critical: the two groups reacted drastically dissimilarly to antibiotics, making it imperative to discover to which group a bacterium belonged before trying to develop a vaccine.

Lamarck on Invertebrates

Lamarck kept working during the Victorian era as well. In 1822, he published his *Natural History of Invertebrates*. In this volume, he made a major contribution to biology when he pointed out the distinction between vertebrates and invertebrates, or animals with backbones and those without.

Taxonomy: Vertebrate

Very much in the spirit of Lamarck 's work was that of Balfour. In 1880, Francis Balfour of Scotland, a zoologist and apostle of Haeckel, would open more fresh avenues in the field of biological classification when he argued that the notochord was not a true backbone, as previous biologists had thought. Instead, he argued that biologists should place animals with a notochord in the phylum Chordata, comprising animals with either a backbone or a notochord, while animals with a true backbone should be placed exclusively in the subphylum Vertebrata taxonomic groupings still extant today. Balfour incorporated many of his observations and ideas in his two-part volume *Comparative Embryology*, which is an extended comparative analysis of both invertebrates and vertebrates.

Circulatory System

Maybe the most solid piece of evidence suggesting that biology was truly transmuting into a science was the experimentation of the German biologist mentioned earlier, Karl Witzenhausen. In 1847, he contrived a prototype of the sphygmomanometer, a device to measure blood pressure. With this he was able to study the circulation of the blood more thoroughly that anyone had previously done. Soon, the Danish physiologist S. A. S. Krogh, a student of Christian Bohr, demonstrated that ultratiny blood vessels called capillaries monitored the flow of blood through the body. For this find, he would garner the Nobel prize in 1920.

Embryology and Reproductive Biology

Another field that began to move more rapidly in this era was reproductive biology. In scrutinizing the female reproductive system, the Berne biologist and author of *De Phaenomeno* Gabriel Valentin and Czech scientist Jan Purkinje found that cilia lining the oviduct cause ova to move through the reproductive system in vertebrates, a finding which the biologist von Siebold later confirmed.

In the closely related field of embryology, the zoologist Karl Gegenbaur of Germany proved beyond all doubt that every cell in the body of any vertebrate emanates from the successive divisions of a fertilized ovum, though no one in Gegenbaur's day fully understood the stages of these divisions blastula, gastrula, and so forth. Within twenty years, the Swiss biologist Rudolf von Kolliker would prove with absolute certainty that the egg is a cell and that all subsequent cells in any biological entity materialize from egg cells.

Rudolf Albert von Kolliker first mastered elementary zoology in. Zurich, the town where he was born, in 1817. The son of an eminent businessman, his first teacher was Lorenz Oken. He later supplemented his studies in Berlin under the tutelage of Johannes Muller. He was on the faculty of Wurzburg from 1847 until his retirement in 1902. He passed away in 1905, the year that Einstein's imperial special theory of relativity appeared. Von Kolliker was responsible as well for adding to both the substance and prestige of still another field of biology which had been stalled for some years histology.

Proceeding relentlessly, Kolliker even faintly anticipated some of the classic conceptions of future genetics as well as cell theory. He prophesied that the cell was the carrier of genetic information, although, of course, he had scant understanding of genes, chromosomes, and so forth. After still more research, he would confirm the notion that axons and dendrites, fibers leading to and from nerve cells, or "neurons," were merely extensions of the basic material of the nerve cell itself. Kolliker then applied this to reproductive cytology (the study of cells) when he showed that both spermatozoa and ova were also cells. In his inspection of the ovum, he subscribed to the notion that since the ovum is also a cell, the behavior of the nucleus, dividing before the entire cell divides, was the critical phase in cell reproduction. The German physician Remak would subsequently fully confirm this view. Former thinkers believed ova were the consequence of fermentation of organic substances. Kolliker was also certainly the first to notice and describe cellular tissue in the so called smooth muscles, such as those lining the intestinal tract. By just past the half-way point of the nineteenth century, Kolliker managed to explain a substantial amount of embryonic development in terms of the successive and progressive division of cells.

Link Between Evolution and Embryology

Just past the midpoint of the nineteenth century, a remarkable link developed between evolution and embryology. The ingenious German biologist Ernst Haeckel appeared with thoughts that would further shake the religious world. He had started his career as a physician but later became a university professor of biology. Though Darwin had his British supporters, Haeckel was among the first reputable scientists in Europe to unflinchingly adopt Darwinism. He introduced his own thoughts on Darwinism in his book *The Riddle of the Universe*.

A principle that caused even greater commotion than Haeckel's adherence to the philosophy of mechanism was embodied in his slogan "ontogeny recapitulates phylogeny"; that is, the developing embryo goes through all of the evolutionary stages that led to the organism in its present form. As he put it in his epic *Generelle Morphologie* of 1866.

Ontogeny is the short and rapid recapitulation of phylogeny, conditioned by the physiological functions of heredity (reproduction) and adaptation (nutrition). The individual repeats during the rapid and short course of its development the most important of the form changes which its ancestors traversed during the long and slow course of their paleontological evolution.

He noted, for instance, that embryos of all vertebrates will have something looking much like gill slits at a given stage in embryonic development. Though hardly anyone takes this to have any profound overtones today, Haeckel redeemed himself a bit when he coined the word "ecology." Part of his belief, too, was the idea that not only did heredity, modify upcoming generations, but the environment did so as well. In this he anticipated what psychologists soon called the "nature nurture" controversy.

9

Embryology and Biochemistry in the Darwinian Generation

Overview

The French chemist Jean Dumas stated for the first time his "law of substitution." This edict stated that under the right conditions of temperature and pressure, a variety of chemicals, including fluorine, chlorine, bromine, and others, could take the place of hydrogen in any organic compound. This made it possible to synthesize, in the ensuing years, an enormous array of other types of materials, including a number that were indispensable to living organisms. An illustration of the phenomenon of hydrogen substitution is the well-known refrigerant freon, found in many freezing units. Chemists construct the freon molecule by substituting fluorine and chlorine atoms for the hydrogen atoms in a molecule of methane gas.

Shortly afterwards, another French scientist, Auguste Laurent, proved that it was thinkable to substitute chlorine for hydrogen in individual substances without altering a substance's properties very much. Since this ran counter to both plain intuition and the attitude of the era, few accepted Laurent's proposals. Later, virtually everyone accepted them.

Two other pioneers in chemistry were the noted Italian Stanislao Cannizzaro and his colleague Avogadro. Avogadro endorsed the proposition that any gas at the same temperature would have the same number of particles in it, whether it was fluorine, oxygen, carbon dioxide, or whatever. Although Avogadro did not know about molecules at this time, Cannizarro submitted some of the earliest evidence that molecules really did exist and were, in fact, the "particles" Avogadro was talking about. Some years after this, Cannizzaro would continue supporting Avogadro's hypotheses, most notably at the International Chemical Congress of 1860.

Another chemical concept vital for biology was the concept of chemical bonding. One of the first to formulate this idea was the Scottish scientist Archibald Couper of the University of Edinburgh, who proposed that carbon chains of indefinite length form the central axis of all organic compounds. He was also the first to publish a structure for aromatic compounds. Although science had to await the likes of G. N. Lewis, Linus Pauling, and Irving Langmuir for any really significant penetration into the mysteries of chemical bonding, the supposition of bonding itself was a major step forward. As it turned out, the bonds in biological molecules such as proteins, amino acids, DNA, and others were put together, to a large extent, with the bonds Couper had discussed.

Emergence of Nutrition as a Science

Equipped with many of these new details, Thomas Osborne of New Haven, Connecticut, showed that there were a huge number of proteins in the body. Beyond this he was the first to discover and recognize the importance of vitamin A in human metabolism, though he did not fully appreciate the far-reaching import of vitamins as an entirely novel class of crucial food elements.

The science of nutrition was making its debut. Further progress would be minuscule for the next forty years. It would not be until 1875 that the Virginia biologist Henry Sherman of Columbia would take another significant step. Sherman managed to show that it is not only the amount of nutrients in the body that is crucial, but the proportions as well. As an example, he proved that the proportion of calcium to phosphorus was as pivotal as the total quantity of each for optimal metabolic function. While at the Carnegie Institution, Sherman also began some inquiries into the amounts of vitamins needed in the diet. Generally, the figures he came up with were too low, and more precise recommendations had to wait, again, until the twentieth century. Still, in honor of his work, he was named president of the American Institute of Nutrition.

Progress in Physiology

Physiology began to grow and divide more sharply than ever during the Victorian era, from about 1840 until the turn of the century. Eventually, there were two discrete fields animal and plant physiology. Earlier physiologists scarcely bothered to try and identify themselves as experts in one field or the other. Now they did so regularly. One of the first topics physiologists surveyed in this epoch was the phenomenon of sensation. The German biologist Ernst Weber, for instance, noted that if he pricked two points on someone's skin with a pin, and if the pricks were close enough together, the person would be unable to discriminate between them. The two pricks would be perceived as a single sensation.

The field had advanced so fast that Johannes Muller felt ready to put together a gigantic tome. In that year he published his Handbook of Physiology, which tallied all of the physiological commentaries of that period. He also championed some ideas of his own, intimating that every nerve has its own "energy," though he was scarcely able to make much sense out of that suggestion. With further exertion, the German physiologist would come a bit closer to modern precepts regarding nerve cells when he demonstrated that nerve fibers emanate directly from nerve cells or "neurons." Interestingly, progress in this field was slow enough that no one felt ready to write a book on the topic of sensation until 1927, when the British neurophysiologist Edgar Adrian book *The Basis of Sensation* appeared.

Not too many years later, the German physiologist Emil Heinrich Bois-Reymond of the University of Berlin opened further windows in physiology. Of French extraction, he began his tutelage in medicine at the University of Berlin, the town in which he was born in 1818. Soon he became a pupil of Johannes Muller, subsequently following Muller as professor of physiology at Berlin in 1855 until his passing in 1896. In this period, he proved using a galvanometer that the nervous system uses electricity to send messages (in "waves," as he called them) to different parts of the body, a precept now a part of the hypothesis behind the transmission of nerve impulses. Most of his work focused on electrical activity in muscles. This fact, understandably enough, tended to further erode the moribund "vitalistic" hallucination that the "essence" of animal existence was inherently undetectable and absolutely immeasurable. In Bois-Reymond 's view, it was the electric current he measured that was the real life-force, or glan vital, so beloved by ancient biologists. He was right in recognizing the importance of electricity in the physiology of living organisms, but he went awry in speculating about the existence of "electric molecules." While the thought isn't altogether preposterous, scientists have, of course, long since forgotten it.

At about the same time, the Scottish physician Sir David Ferrie conducted some monumental studies on the brains and nervous systems of a range of animals. His goal was to work out the locations of motor nerve impulses leading to muscle motion and sensory nerve impulses, which carried outside stimuli to the brain and nervous system. Still more gains came when, in 1850, the German physicist Gustav Fechner submitted that there is no simple linear relationship between the strength of a sensation and the intensity of the stimulus. Rather, as the strength of the stimulus increased, the sensation would increase far faster.

The German physiologist Hermann von Helmholtz added to the conclusions of the fabled eighteenth-century Italian physiologist Galvani. Hermann Ludwig von Helmholtz is unique in the archives of science in that he was one of exceptionally few who made important contributions to both physics and biology. He was born in 1821 near Potsdam, the child of a teacher in the local gymnasium school. He apprenticed with Johannes Muller in Berlin, and soon obtained his medical degree. In the years to come he served as a military physician and eventually became a member of the faculty at Koningsberg, the home of the extraordinary philosopher Immanuel Kant. Ultimately, he became professor of physics at Berlin and at the new physico-technical institute in Charlottenburg. He held these positions until his demise in 1894. Although he took notice of such fields as biology, philosophy, physics, and mathematics, he accomplished nothing of consequence in philosophy and mathematics.

One of his choicest contributions complemented Galvani's findings. He was the first to measure the velocity of a "message" along a nerve fiber. He accomplished this by paying close attention to the neurons of frogs. He discovered that such impulses travel at about thirty meters per second, and he showed further that chemical changes are critical to the health of the nervous system. He would later go on and apply this knowledge to the study of the sense organs.

Helmholtz and the Ear

In 1856, Helmholtz proposed the "resonance" principle of hearing, when he suggested that membranes in the cochlea of the inner ear act as resonators to amplify sound. Understanding of the hearing mechanism took another surge forward in 1863, when Helmholtz supported the assumption that via a set of resonators on the cochlea, the ear determines the pitch of the sound an animal hears. Yet it would not be until the 1950s that physiologists, taking the right clues from Helmholtz 's findings, could claim that they had indeed grasped the mechanism of hearing. In 1961, the Hungarian physicist Georg von Bekesy captured the Nobel prize for his work in this area. In his theory, "waves" that appear on the fluid in the cochlea carry the external vibrations which are then processed by the brain to produce hearing.

Bell and Neurophysiology

Evidently, the understanding of physiology was advancing rapidly, and Charles Bell explored this field still further. Born in 1774 in Edinburgh, the son of a preacher, he spent his formative years in the worst sort of poverty imaginable. His father was barely able to feed the family from one day to the next. Even so, he was able to enroll at the University of Edinburgh, eventually getting his medical degree.

He later became a curator of the hallowed Hunter Museum and eventually a professor of anatomy in Scotland. In 1830, he penned *The Nervous System of the Human Body*, in which he distinguished a miscellany of separate types of nerves including sensory and motor nerves, a find later confirmed by Magendie. In "Idea of a New Anatomy of the Brain," a short paper which he wrote in 1810, he described the effects of severing the medullary nerve. He had discovered that the medullary nerve functions in two ways, depending on whether the experimenter tampers with the anterior or posterior root. If he touched the posterior root, the muscle connected to it could not contract. If, on the other hand, he touched the anterior root, the muscle would go into spasm. This explained a problem no previous biologist had been able to explain how it was that nerves could both send and receive nerve impulses. Although Bell's explanation is far from complete, he did apparently understand that the double roots made the double function possible, adding still more data to the growing fund of information about muscle anatomy and physiology. He, like so many others before and after, also tried to localize "mental states" in various parts of the brain in the spirit of modern inquiry into the philosophy of mind. In this manner, he exerted considerable leverage over recent academic surveys of the nervous system as well as on speculations about the constitution of the human mind that are in vogue today.

Helmholtz and Botany

Like Purkinje, Helmholtz became a most acclaimed and popular teacher as well as researcher of his generation. Among his research contributions are experiments, performed in 1862, in which he coated plant leaves with wax and exposed them to sunlight. He promptly realized that only the plants with uncoated leaves were manufacturing starch, a clue that sunlight was necessary for photosynthesis. Both German botanists Julius von Sachs and Nathanael Pringsheim would confirm that chlorophyll exists in the plastids of plants and is not diffused throughout the tissues.

By 1840, the French botanist Jean Boussingault swelled the already large corpus of facts on plant physiology when he proved that the amount of nitrogen and carbon in plants could not be explained merely by referring to the sum total of these elements in manure, which is a source of carbon, hydrogen, oxygen, and nitrogen. He showed that plants obtained the nitrogen they needed for maturation from nitrates found in the soil. He also demonstrated that carbon came directly from the carbon dioxide in the air.

Rise of Biochemistry and Some Practical Applications

This progress in botany led others to search for practical applications which, in turn, yielded still more knowledge of biochemical processes. The German biologist Justus von Liebig, in his *Applications of Organic Chemistry to Agriculture and Physiology*, believably started the modern branch of biology known as biochemistry. Born in 1803, he was the son of a Darmstadt merchant. His initial fascination with chemistry stemmed from helping in his father's shop. He considered becoming a pharmacist for a while, but, ultimately, in the laboratory of the eminent chemist Gay-Lussac in Paris, he became completely absorbed in chemistry. Eventually, the University of Giessen appointed him to the faculty as professor of chemistry. Arguably, the creation of biochemistry in Germany and even much of Europe was due to his efforts. His principal interests were in organic chemistry, and he contributed much to the techniques for analyzing organic compounds.

He did the spadework for biochemistry when he explained how flora and fauna swap carbon for nitrogen, thereby taking a major step toward comprehending the nitrogen cycle the process whereby green plants convert nitrogen in the air and in soil into material they can use. He accomplished this when he began his appraisal of manure, showing that plants derived their carbon from carbonic acid and that ammonia served as their principal source of nitrogen. This research constituted a major contribution to plant physiology as well.

His inquiries were not without either inaccuracies or criticism. Among others, the nineteenth-century German biologist Matthias Schleiden criticized him, particularly for denying that manure was a source of nitrogen for plants. Criticism also descended upon him for ignoring the role of plant respiration and for his odd fantasy that crops would thrive better if farmers supplied them with insoluble phosphoric acid. Current agricultural science now knows this is quite wrong. It is vital to grasp, nonetheless, that Liebig evidently brought much of this opprobrium upon himself because of his overbearing and critical demeanor he was very likely one of the most unpleasant biologists in history.

The American biologist Samuel Dan would later extend biological comprehension of the role of manure by revealing that phosphates are the primary compounds in manure that act as a fertilizer of flora.

Embryology of Von Baer

Then came the important work of the Russian biologist von Baer in the field of embryology. Karl Ernst von Baer first saw the light of day in 1792, in the modern state of Estonia, though he was of German lineage and German nobility. This status aided him immensely in his later life. He first attended school at Revel, where he began his education in natural science. In 1810, he began studying medicine at the University of Dorpat. He says in his *Autobiography* that he would have liked to have studied biology as such, rather than medicine, but that, in those days, the only route to science was via medicine. This remained true until late in the nineteenth century. Many of his early quests were in epidemiology, or the science of the spread of disease, where he focused on the people of his own country.

After leaving medical school he headed for Wurzburg, where he began investigating comparative anatomy. He purchased a leech for his first dissection. The preeminent faculty member at Wurzburg during this period was Ignaz Dollinger, a former apprentice of the German philosopher Schelling. Dollinger, with his combined ardor for philosophy and anatomy, managed to instill a lasting interest in the origins of human life and the evolution of animal species in von Baer.

At the university, von Baer also began to read widely in physiology and anatomy. However, money rapidly became a problem and he traveled to Berlin in hopes of creating some kind of medical practice to bring in badly needed funds. Inevitably, his talent became surpassingly clear, and the University of Konigsberg appointed him professor of biology, charged with taking care of the university museum the perfect environment to support his own interests. After a series of brilliant studies, he became dean of the medical school and then rector of the entire university. In 1827, he found that all mammals originate from eggs stored in the ovaries of the females of the species. Much of this research appeared in his 1828 book the *Epistele*, in which he described his discovery of the mammalian egg.

He followed this volume with *The Developmental History of Animals*, which subsequently became a standard embryological text. In this, a work of vast importance, he recounts his finding of the ovum as well as the notochord a primitive structure seen only in the embryos of vertebrates which gives rise to the vertebral column. Also, he cautiously set out his "germ-layer" theory of embryonic development. According to this conjecture, the fertilized egg in mammals develops into four distinct layers of tissue (later corrected to three by the German physician Remak). Out of each, distinct organ systems would develop.

The Biogenetic Law

This was the first appearance also of what biology now calls the "biogenetic law," the belief that a higher animal passes through states that resemble stages in the development of lower animals. More accurately, this law says that the general characters of an organism appear before the more specialized ones. Also, the specific and detailed features develop out of the general ones. Although von Baer did more or less cogently describe these modifications, he had no notion of evolution and made no attempt to relate these findings to evolution.

Von Baer also aided the evolution of embryology as a science in more sweeping ways. For one thing, he transformed embryology, following the lead of anatomy, into a "comparative" science by emphasizing the advantages of comparing the embryological development of various species with one another. He also offered some of the earliest explanations of the development of the amniotic sac and the urogenital system, as well as the digestive and nervous systems.

Von Baer was also mesmerized by nature and the wilderness. At various times he headed expeditions to Lapland and Nova Zemlya and sailed the Caspian Sea to research marine flora and fauna. In his last years he founded the Society of Geography and Ethnology at St. Petersburg and supported the launching of the German Anthropological Society. He received many honors throughout his career, including the prestigious Copley medal as well as an appointment to the Paris Academy of Sciences. His influence on future biology is unquestionable. In modified form, his germ-layer concept still exists today. Beyond this he was something of a minor prophet. He predicted, among other things, that a comparative anatomical analysis of the great variety of distinct organs in the animal kingdom would be a critically important investigative implement. This prophecy has long since been fulfilled.

Research Work in Embryology

The German physician Robert Remak would bring all of this to the highest level of perfection. Born in Posen in 1815, Remak was the offspring of a Jewish family. He studied biology under Muller, who, in recognition of his talent, aided him in achieving faculty rank. Nonetheless, Remak earned his keep for most of his life as a medical doctor.

Remak found some comparatively minor blemishes in von Baer's work when he noted that there were actually only three, rather than four, primitive layers of tissue the ectoderm, mesoderm, and endoderm. The general idea and other details of von Baer's conclusions turned out to be approximately valid as stated before. Remak also recast some of Schleiden's research, showing, in 1841, and contrary to Schwann's conviction, that all cells came from previously existing cells. In the process of attaining this he scrutinized the development of frogs' eggs, proving quickly that the egg is itself a cell which continues to divide, forming new "daughter" cells. He testified too that cell division begins in the nucleus. He bolstered this work with further studies of the comparative embryological development of both frogs' and birds' eggs. He then coined the terms "holoblastic," to refer to eggs that divide into two equal parts typical of frogs' eggs and "meroblastic," whereby only partial division of the egg occurs, resulting in two unequal parts characterizing birds' eggs. It was at this point that he corrected von Baer's work and thereby distinguished the now-familiar three germ layers in the embryo. The outer, or ectoderm, he hypothesized gave rise to the skin and the nervous system, the middle, or "mesoderm," to the musculature, and the inner, or "endoderm," to the digestive tract. We know today that all of this is roughly correct.

Embryology of Invertebrates

After this work, the Russian biologist Alexander Kovalevski of the University of St. Petersburg observed other vertebrates and invertebrates, quickly finding that von Baer's findings applied to invertebrates as well. Thus, even invertebrates began the life cycle from the three basic layers in the embryo. Kovalevski would also do a considerable amount of work on the notochord, or backbone, as well as promote Darwinism in Russia.

Refinements of Von Baer

Yet, still more probings into von Baer's results were imminent. Toward the middle of the nineteenth century, the German physiologist Johannes Muller made his claim to everlasting renown. The son of a comparatively affluent shoemaker, he started his schooling at the University of Bonn, a moderate distance from the town of Coblenz on the Rhine where he was born in 1801. After obtaining his medical degree, he headed for Berlin, and in 1830 became a member of the faculty at the University of Berlin.

Historians universally embrace him as one of the great physiologists of his day. Among his many publications is the *Handbook of Physiology*, mentioned earlier, published in 1840. Many scientists and scholars used this tract very widely during the nineteenth century. His concerns were sweeping, extending from physiology and chemistry to comparative anatomy and physics. Muller refined von Baer's ideas, further describing as well as emphasizing the existence of the three basic layers of embryonic tissue mentioned above. He was responsible too for his doctrine of "specific nerve energies," a somewhat contentious belief at the time which stated that nerve impulses make their debut in the sense organs and then travel through the central nervous system. In general, he added much to biological comprehension of how the immediate environment affects the nervous system and, of course, the whole organism.

More enigmatic was his *Handbook of Human Physiology*, in which he tried to explain human thought "mechanistically." Again, this approach ran into extreme opposition, primarily because such a view was inconsistent with the belief in the human soul as something separate and distinct from the body and which is the locus of thinking. Another line that branched off of this was the odd notion that the mind is a "computer" and that all mental states are really "computational" states. (In the twentieth century, all of these versions of "mind-brain materialism" appeared with reckless abandon. From the 1960s on, nonetheless, many philosophers, inspired by the writings of the Austrian theoretician Ludwig Wittgenstein, cast considerable doubt on all these propositions). As Muller's career neared its finale, he trained a number of followers who carried on his work, among them Theodor Schwann, who formulated the cell theory, and the German biologist Rudolf Virchow, who conducted important investigations in pathology.

Muller spent his last years plagued by, ironically, the very phenomenon that had catapulted him to stardom nerves. He had suffered from depression since childhood. The debilitating effects of age, plus the trauma of suffering through a shipwreck, inevitably overcame him. After a certain point he was never able to rebound from his melancholia. Although no one has ever proven it, his own recognition of his waning powers plus his inherent melancholia make it plausible that he committed suicide. He was simply found dead in his bed one morning, although he had not been suffering from any apparent illness. How exactly he killed himself if indeed he did do so is not known.

Advances in Technology

Others were improving the microscope still further. In 1830, Joseph Jackson developed an "achromatic" lens, a lens made of two or more materials so that different colors focused at the same location, thereby avoiding distortion of the image. Although such lenses already existed, no one had fabricated any expressly for the microscope. Toward the close of the century, microscopes with such achromatic lenses were growing common, avoiding the devilish problem of chromatic aberration that had haunted the older microscopists. Before long, many universities and museums as well as wealthy amateurs were able to purchase these spectacularly altered microscopes. By 1840, the popularity of the microscope had swelled so much that the Italian physicist Giovanni Amici was also busily improving the relevant techniques. He was able to devise a so-called oil-immersion microscope which, because of the optical properties of oil droplets, had an enlarging power of 6, 000X far greater than anything previously. With this microscope he became the first scientist to actually follow the process of plant fertilization.

Soon, the amateur microscopist John Dolland appeared. A novice of Kingenstierna 's, he was a brilliant mathematician and physicist. In 1844, he fulfilled an ambition by constructing the so-called waterimmersion lens; that is, he used water as a medium to separate the lens and the glass slide. After his work came the contemporary microscope, with the ocular eyepieces and the objective separated by a movable tube, the variety familiar to college undergraduates everywhere.

Rise of Cytology

With the aid of these magnificent technological refinements, the inspired Italian biologist Camillo Golgi rapidly entered the front ranks of the world's great biologists. Born in 1844, he got his medical degree at the University of Padua in Italy. Then, in 1875, he became professor of histology at Padua. Among other things, he streamlined scientific investigations of the nervous system and cytology by devising a technique for staining cells using silver salts, which allowed the investigator to see the cells much more easily. With these tools, he became the first to discover the synapse the gap between nerve fibers. In a word, Golgi showed that nerve fibers do not constitute a continuous thread but are, instead, separated by these synapses. He ventured into the field of immunology as well, examining a large number of malarial parasites.

Much later, in 1883, he found a type of cell organelle essential to the functioning of the nervous system. He confirmed and elaborated on his discovery in 1898 when he described in detail this structure found in the cytoplasm of cells. Subsequent generations of biologists named these organelles "Golgi bodies," in his honor. In his essay "What Is Its [cytoplasm] Structure, and How Does It Work?" Professor John E. Harris describes the Golgi apparatus:

> A more elusive cytoplasmic inclusion is the so-called Golgi apparatus, a complex structure of water vacuoles partially sheathed with fatty material and revealed in a protean diversity of form by various complicated histological methods.

Golgi would enrich the world still more with his research on malaria and pellagra. His toils revealed that of the two brands of malaria, so-called intermittent and pernicious, the former develops when parasites invade only the blood; but when they invade the brain and other organs, the pernicious form of malaria appears. He showed also that the severity of the disease varied directly with the number of parasites in the blood. Golgi was predestined to win the 1906 Nobel prize for his research on the nervous system and olfactory system, as well as for distinguishing between a variety of different nerve cells.

Pfeffer and Semipermeability

Related to this was the work of the German botanist Wilhelm Pfeffer. Pfeffer had begun, in 1878, his epochal probing into "semipermeable" membranes, which are tissues that allow only molecules below a certain size to pass through them. He used this concept to ascertain the weight of a single protein molecule. However, further insights into the detailed architectural structure of proteins would have to wait until the twentieth century with the offerings of thinkers like Dorothy Wrinch, Irving Langmuir, Erwin Chargaff, and several others.

Study of Disease and Blood

The German physician Paul Ehrlich would push ever deeper into the nature of disease. Ehrlich is, defensibly, one of the initial heroes in the fields of hematology the study of blood formation and function as well as chemotherapy. He was born in Strehlen in upper Silesia to a Jewish family. He became enthralled with chemistry in childhood, but his attention soon turned to biology. He received his medical degree from Leipzig in 1878. Even as a undergraduate he began investigating new frontiers. Using a type of dye called an aniline dye, he found and categorized nearly every discrete difference among white blood cells. By 1881, he was using the more efficient dye methylene blue. Its effect on some bacteria provided him with a clue that this dye might be medically useful in combating certain bacterial infections. In fact, he found the dye methylene blue was effective against malaria. He then generalized his reasoning as follows: if certain tissues were selectively affected by certain dyes, perhaps other tissues would also be selectively affected by different drugs. He had hit on the concept of the "magic bullet" a drug that would selectively destroy certain invading organisms while leaving the host tissue unharmed.

His work on dyes eventually led him to experiment with other chemical compounds. By tinkering with various drugs previously found ineffective against the disease trypanosomiasis, he found a new and colossal tool in the war on syphilis. After years of struggle, pharmaceutical companies marketed his innovation as Salvarsan, although since the advent of penicillin physicians rarely use Salvarsan. For these efforts, Ehrlich would receive the 1908 Nobel prize.

In 1894, the University of Tokyo pathologist and student of Koch, Shibasaburo Kitasato and the French bacteriologist Alexandre Yersin of the Pasteur Institute would each discover the bacterium answerable for bubonic plague, that infamous scourge of civilization that wreaked havoc in ancient Rome, the late Middle Ages, and other periods of the past. Kitasatc, also probed such scourges as anthrax, tetanus and diphtheria.

Just before the turn of the century, the French physician PaulLouis Simond discovered that rat fleas are the vehicle of transmission of many illnesses to human beings, a discovery that emerged while he was fighting the plague in India. In the twentieth century, of course, other scientists would follow similar paths and develop "vaccines" for an assortment of animal and human disorders, including anthrax, polio, smallpox, yellow fever, diphtheria, and many others. Paul Ehrlich, as suggested above, all but created the field of immunology when he perfected a vaccine against diphtheria.

Subsequently, in 1913, Bela Schick devised the now well-known Schick test for diphtheria, allowing physicians to quickly and easily diagnose the dreaded ailment. The beginning of the end for yellow fever as a slayer of millions came in 1900, when Walter Reed began his probings into the epidemic of that malady in Cuba, where he and his team found that the *Aëdes* mosquito (more accurately, the *Stegomyian* mosquito, a subgenus of genus *Aëdes*) transmitted yellow fever.

Conclusion

In sum, there was tremendous progress in chemistry with the work of Dumas, Couper, and Cannizzaro, as well as Avogadro's famous research. Physiology also made gains with Weber's and Muller's work on sensation. In the adolescent field of cytology, the legendary Camillo Golgi made his momentous discovery of Golgi bodies. Medicine, too, was rapidly moving toward the twentieth century, as Ehrlich found his "magic bullet," while Kitasato and

Yersin solved the mystery of bubonic plague and Bela Schick developed the Schick test for diphtheria. No one was overlooking technological advances either, as Amici developed the "oil-immersion" microscope while Klingenstierna nurtured the water-immersion lens.

10

Genetic Engineering and its Evolution

Introduction

The American biologists Charles, Sibley and Jon Ahlquist of Yale in 1984 applied DNA experimental techniques to reveal that the relationship between humans and chimpanzees was much closer than the relationships between chimpanzees or humans to other great apes. They also devised an evolutionary tree for birds in their 1986 "Scientific American" article, "Recasting bird phylogeny by comparing DNA's." In a similar way, they also applied DNA techniques to show that previous descriptions of how and from what species songbirds and other birds had evolved was not accurate. The year 1983 also saw much genetic research. The American biologist Walther Gehring and his collaborators found the so-called homeobox a particular sequence that coordinates the general development of certain structures in mammals and segmented worms. Also in 1983, the American biologist Barbara McClintock captured the Nobel prize for her breakthroughs on the genetics of corn, which had almost at once led to spectacular increases in corn farming efficiency. At about the same time Andrew Murray and Jack Szostak oversaw the creation of the first manufactured chromosome.

Genetics in the Service of Medicine

Science matched this exploit in 1990, when biologists found the gene responsible for neurofibromatosis, commonly known as "elephant man's disease" because of the gross disfigurement it causes. That disclosure led them to the protein the gene produced. On July 31, 1990, the Recombinant DNA Advisory Committee sanctioned the first successful gene therapies for human beings. These included treatment for such a malady as adenosine deaminase deficiency, a genetic ailment that quickly obliterates the immune system.

Rise of Bioethics

In the 1990s biology witnessed the emergence of a entirely new field of biology and philosophy bioethics. Biologists, philosophers, theologians, and scientists joined forces to probe the myriad ethical problems involved in biology and medicine. Signs of growing interest in the new field had appeared in the 1970s and 80s with the advent of recombinant DNA and other scary medical/biological technologies. Also in the 1970s, environmentalists had begun voicing mounting agitation over population growth and the destruction of the environment, whether through the dumping of chemical wastes or through nuclear pollution. In 1972, Harvard University sponsored one of the very first conferences that devoted a substantial amount of time to the ethical aspects of biological questions. In 1976, the proceedings of the conference were published by D.C. Heath/ Lexington Books as *Issues in Population Education*. By the 1980s ethical concerns had expanded to encompass new horrors over the possibility that impetuous genetic technological orchestrations might inadvertently unleash a devastating new organism. Movies like The Andromeda Strain from 1971 only served to heighten such alarm. Soon, the assault on science broadened.

Political developments will affect the course of ethics. A recent example is the appointment of John Gibbons, previously the director of the Office of Technology Assessment for thirteen years, as President Clinton's chief science advisor. There is powerful evidence that Gibbons has a great deal of sympathy for animal rights groups that have often trashed scientific laboratories using animals for research. An aide to Gibbons confirmed in February of 1993 that Gibbons's wife was a member of PETA (People for the Ethical Treatment of Animals) and that Gibbons himself had supported PETA 's attempts to secure their overall goal.

Science itself became the target of charges of fraud, which included allegations of scientists manipulating data. The years following showed that, in fact, corruption and fraud was and is a dark, unchecked, and pervasive malignancy in the scientific community. One of the most recent developments concerned the so-called David Baltimore case. In 1986, Professor Thereza Imanishi-Kari, then on the faculty of Tufts University, published a paper in the journal Cell claiming that a gene transfer into mice had had a far greater affect on their immune systems that anyone had thought it could. Baltimore was a co-author of the paper, although he was not accused of any wrongdoing. Instead, critics charged that Professor Imanishi-Kari had not in fact conducted some of the experimental trials she claimed she had done. Although her guilt or innocence has not been conclusively established, subsequent evidence appears to have cleared her, and July oil 1992, the U.S. Attorney's Office decided there was insufficient evidence to convict her. Still, the fact that such allegations keep appearing illustrates the degree of concern over the problem of scientific fraud.

Ethical Problems in Medicine

Ethical qualms escalated with the advance of technology, as opportunities opened up for experimenting with embryos, using in vitro fertilization techniques to produce pregnancies for couples that might otherwise go childless, and so forth. So it was that in 1978 the first "test-tube" baby made her debut. A product of artificial insemination of the egg outside of the mother's body, her birth merely added fuel to the burgeoning controversy surrounding all such manipulations of the ordinary reproductive processes. The abortion controversy deepened and heightened trepidation over ethical aspects of many biological and medical procedures. Many became increasingly disturbed over the use of laboratory animals in medical experimentation. Surrogate parenting became an emotionally charged issue after the notorious "Baby M" case.

The very definition of "death" became a mammoth ethical problem. With the advent of imaginative technologies that allowed the brain-dead, or human beings in "pvs" (persistent vegetative state), to remain alive, even after their brains had ceased functioning, many physicians and ethicists counseled a radical modification of the definition of death. Instead of the old criterion whereby a physician could declare someone dead after the heart and lungs had stopped functioning, many now proposed that physicians should be able to declare someone legally dead when the entire brain had ceased functioning, even though machines could keep that person breathing. With such suggestions came the complex problem of euthanasia. Could it ever be ethically justifiable to end someone's life when they were still breathing, although any potential for meaningful existence had vanished? Could such actions be morally forgivable when a person was sentient, but suffering from an incurable and devastatingly painful pestilence, such as cancer or AIDS? The issue has become further tangled very recently with research appearing in various journals offering powerful reasons for believing that the concept of brain death is radically confused.

Positively the most bizarre idea in recent history is cryonics. If dying of an incurable disease, you can, for a huge fee, arrange for scientists to freeze your head, or your entire body even more expensive in liquid nitrogen. The theory is that after they have "popsickled" you for a few generations, some new technological marvel of the next millennium will allow you to be thawed, cured, and to get on with your life. Predictably, this eccentric theory is extremely controversial, although it is difficult to prove that it is technologically futile. Of course, ethical and spiritual difficulties abound here as well, in such efforts to cheat the crypt.

Human Genome Project

As always, scientists proceeded anyway. At the genetic level, again, there were a number of innovative technological advances and new projects. For example, there is the Human Genome Project, which has carried with it the fear that eugenics, or improving the race via genetic tinkering, might develop into an attractive philosophy to some. Biologist Robert Sinsheimer, reknowned for discovering a single-strand DNA and participating in the in vitro replication of infective DNA, then Chancellor of the University of California first instigated the recommendation in 1984. Reflecting on the large and costly requirements of biological investigations, it occurred to him that it would be enormously useful to map all of the three billion bases that constitute human chromosomes. Such a complete genetic map of human beings would make the search for defective genes that cause certain diseases much easier. A good example of this was the search. by geneticist Raymond White of the Howard Hughes Medical Institute of the University of Utah. Although he had, by 1985, by studying Mormans of Utah and southern Idaho narrowed the search for the gene causing cystic fibrosis to a narrow region of the chromosome, the gene was not found until 1989. Many scientists feel that if White had had the complete "map," the gene would have been found much earlier. The United States Department of Energy began to fund the project, since the scientific community already knew that radiation caused genetic damage. Later, both the National Research Council and the National Institutes of Health added their forces to the project.

Perhaps most importantly, James Watson, co-winner of the Nobel prize for unraveling the structure of DNA, became the project head. Under Watson's leadership the project devised new ways of attacking the problem. One was to construct a "map" of all chromosomes by finding a "marker," or short piece of DNA that scientists can recognize indirectly, since they already know the appearance of adjacent sections. Such markers occurred about every two million bases or so.

Eventually, both the European scientific community and Japan joined in the project with geneticists everywhere, spotlighting some very specific aspects of the problem. As mentioned already, the project is not without somber critics. As always, there are concerns. Many brood over the possibility that unscrupulous people might use the facts gained through the project to weed out genetic "undesirables." Such fears notwithstanding, if all goes well, this genetic information could facilitate the determination of the evolutionary ancestry of many species of animals. Scientists may be able to tell, for instance, whether it is really true that human beings are the descendants of humanlike creatures who lived in Africa about a quarter of a million years ago, though at the moment this is still extremely debatable.

New technology will certainly figure into this. One of the most promising recent developments was the creation of the nation's first department of molecular biotechnology at the University of

Washington in 1992. Its head is the brilliant and energetic Leroy Hood, who perhaps looks more like a movie star than a scientist. With better computers and even robots, he believes that the manipulation of genetic data will dramatically accelerate biological progress. If all goes well, the department expects to use its technology to finally produce a complete map of the hundred-thousand genes and the three billion nucleotides, or bases, that comprise them. That, in turn, should allow the medical profession to do such things as predict the health a given person might expect to enjoy during his or her lifetime. As always, money is an issue and many scientists have suggested that the money ought to be spent elsewhere.

An interesting recent offshoot of this is the "dog genome project." Professor Jasper Rine, a geneticist of the University of California at Berkeley, is currently trying to map the entire dog genome. If successful, this project will allow Rine and others to find the genes that cause many sorts of dog behavior, including such highly specific traits as the tendency of Newfoundlands to rescue drowning people. As Rine explained in a 1993 interview,

> *"A major contribution of mouse genetics to human genetics has been being able to map genes from one organism so one would know where to look for similar genes in another organism having a third mammal with a high-resolution map will aid this comparative mapping."*

Closely related to the Human Genome Project is an old idea of the biologist Sydney Brenner of the Medical Research Council laboratories in Cambridge the odd-sounding notion that it might be possible as well as useful to know everything there is to know about comparatively crude organisms, for reasons similar to those stated by Rine above. Brenner had recommended Caenorhabditis elegans for such a program, since in this and certain other nematodes (a variety of worm), there are less than 1,000 cells. An assault on this worm did in fact begin and science today often calls the enterprise "The Worm Project." In fact, there have been bonanzas. Biologists know the embryonic lineage of each of its cells, what the ancestral zygote was like, and so forth. They accomplished all of this in less than twenty years. By 1986, biologists understood thoroughly this worm's nervous system and were well on their way to constructing a "genome," or a genetic map, of this organism. Scientists envision this as a preliminary step for the Human Genome Project.

But while molecular biology has undeniably been the most bewitching sort of inquest going on in recent history, classical biology, or the biology of the entire organism, is thriving too. Among other things, paleontologists were unearthing many undiscovered species of both plants and animals. By the end of the 1980s, for instance, biologists had found two new species of lemurs, never before seen, on Madagascar, as well as an atypical species of monkey in Brazil. Most recently, a creature was found in Pakistan having both goat and cow features.

New Technology in the Service of Fossil Studies

New technology also revived old conundrums in paleontology, such as how the dinosaurs had vanished. Using new techniques at first intended for physicians, such as CAT scans computerized axial tomography paleontologists began scrutinizing fossil bones. Not surprisingly, much controversy erupted. In 1989, some scientists found hollow spaces in museum dinosaur bones. Since bird bones are hollow, they theorized that the relationship of birds to dinosaurs may be much closer than anyone had suspected. Other inquiries focused on comparatively less global issues and more on matters of detail. Examinations of *Tyrannosaurus rex,* for instance, indicated that their front limbs were far stronger than anyone had previously envisaged. Yet the most amazing find occurred in 1989, when the paleontologists Philip Gingerich of the University of Michigan Museum of Paleontology, Holly Smith, and W. L. Simons chanced upon fossils in Egypt intimating that there once lived a whale with hind legs. Supposedly, this fifty-million-year-old creature was still an ocean animal as the hind legs appeared too weak and puny to carry it about on land. One speculation was that they had been useful in reproduction.

Embryological Advance

In the field of embryology, the cutting edge has, for some years, been on "feedback" mechanisms. The idea here is to explain how a fertilized egg develops the various kinds of tissues that constitute the adult organism. In the feedback thesis, a cell has something like an ordinary thermostat. When the temperature within the cell reaches a prearranged level, the "furnace" shuts off. When it reaches another level, the cell starts to form, say, nervous tissue. Another view proposes that certain hormones called "histones" may either engender or retard certain sorts of tissue synthesis. When the egg reaches a certain stage of embryonic development, a particular histone appears which "tells" the cell to start producing, say, striated muscle tissue.

Epilogue

Where biological curiosity will lead in the years to come is anyone's guess. It is, I think, inevitable that someone, someday, will clone a human being. Should that happen, our traditional concepts of "personhood," the soul, and much of our ethical tradition will behold itself in turmoil. That said, it is an unarguable truth that the field of bioethics will simply continue to expand as more and more troubled scientists, philosophers, theologians, and laypeople try to meet such crises.

In geology a spectacular claim emerged in January of 1993, adding to the continuing controversy over "catastrophism." At the annual meeting of the American Geophysical Union, the geologists Michael Rampino of New York University and Verne Oberbeck of NASA's Ames Research Center in California claimed that asteroids colliding with the earth might have initiated movements of the continents, producing debris long supposed to have been caused by glacial movement. Such collisions, some thought to have originated some 250 million years ago, could have triggered the mass extinctions the are thought to have occurred at that time. According to some estimates, close to 96 percent of all species perished in that cataclysm.

In paleontology, a remarkable find also occurred in 1993, when Paul Sereno of the University of Chicago found a complete skeleton of the most primitive dinosaur so far discovered. This adds tremendously to our knowledge of the evolution of dinosaurs. In the words of Professor Sereno, the new dinosaur "is close to what we expected the common ancestor of all the dinosaurs to look like."

Genetic engineering will most likely keep scaring many in the very understandable way that it has so far. Apprehension about doomsday viruses will not easily disappear. If we can responsibly overcome such problems, both ethical and scientific, the future progress of biology and the potential for the genetic treatment of diseases is boundless.

11

Chemical Revolution

Introduction

Its drama lay in a conflict that split the chemical community into opposing factions but its significance is more complex. According to the most popular interpretation, the chemical revolution consisted in the overthrow of the phlogiston theory and its replacement by the oxygen theory of combustion. On a grander scale, the episode has been hailed as the founding era of modern chemistry and Antoine Laurent Lavoisier (1743–94), creator of the oxygen theory, as its 'father'. In recent decades, the conventional view has been challenged. In the first place, combustion was not an overwhelming concern of eighteenth-century chemists, so it is not clear why a new combustion theory should provoke such strong reaction, nor be equated with the ascendance of modern chemistry. Historians now ask what the chemical revolution was really about. Secondly, Lavoisier 's contributions have often been studied out of context, with the result that continuity of chemical inquiry has been overshadowed and the epistemic break with tradition possibly exaggerated. Current investigations aim to relate Lavoisier's achievements to the work of his predecessors and contemporaries. Because the chemical revolution is a major episode in the history of science, often cited by historians, philosophers and sociologists to illustrate their conceptions of scientific change, these problems have implications that transcend the episode itself.

Chemistry in 1800 was certainly different from what it had been a century earlier. Proponents of Lavoisian chemistry shared a strong conviction that its adoption represented a rupture with the past, as the words of a young Swiss convert, Marc-Auguste Pictet (1752–1825) illustrate.

Chemistry, banished until now to a small circle of adepts whose language and ideas were equally obscure, has become the inseparable aide and companion of Physics: these sciences, united and guided by experiment alone, have proceeded at a rapid pace; Chemistry itself has undergone a great revolution, a frightful scaffolding has given way to a simple and illuminating theory, based upon the immediate consequences of experiment. Logic, without which experiments are mere isolated facts, has become more severe in its conduct and more certain in its results: in short, everything indicates that we are on the right path, and that it will lead daily to discoveries in the natural sciences.

Pictet 's view was from the vantage point of a triumphant campaign for the new theory. Chemistry, in the eyes of its promoters, had been created anew with a methodological soundness and a theoretical rigour which was lacking before. But to what extent were its advances due to the Lavoisian innovations rather than a century of evolutionary development?

For an appraisal of the status of chemistry prior to the upheaval, we may turn to Gabriel Venel's article 'Chymie', written at mid-century for Diderot's *Encyclopedie.* Venel (1723–75) found that chemistry in France suffered from a two-fold sense of inferiority. On the one hand, it was not perceived by French intellectuals as a science on a par with mathematical physics. Its image was still tarnished by association with alchemical 'puffers'; it seemed to lack a powerful organising theory. Secondly, French chemistry had not enjoyed the encouragement accorded to its German counterpart. Although critical of the current state of chemistry, Venel was sanguine about its potential. To make chemistry respectable required a breakthrough to an overview which would give coherence to chemical doctrine. Venel called for a 'new Paracelsus' with the insight and the dramatic flair necessary to bring about a 'revolution' in the science.

Developments in the first half of the century gave Venel grounds for optimism, for there had been notable advances in chemical practice and theory. Practical chemistry contributed improvements in mineral analysis and metallurgical techniques, as well as isolation of new materials, including phosphorus. Nor did chemistry lack for theory; several frameworks were available, including the venerable theory of the four elements, the doctrine of essential chemical principles and the notion of chemical affinities. Problems arose in the compatibility of diverse theories, or, more pressingly, in the relation of theory to laboratory practice. Venel's discontent with the chemical literature was twofold. On the one hand he found a plethora of petty chemical treatises competently treating matters of detail but supplying little interpretation. At the other extreme were speculative treatises, written by philosophers with little chemical knowledge and contributing little to advance the science. Venel espoused a 'philosophical chemistry' that would follow an intermediate path, guided by an intuitive method grounded in laboratory experience rather than extrinsic philosophical doctrine.

Most French chemists at mid-century accepted that natural bodies were composed ultimately of earth, water, air and fire. The conviction had been reinforced by Stephen Hales's discovery in the 1720s that air could be extracted from animal, vegetable or mineral substances and trapped (by displacement of water) in a 'pneumatic apparatus'. In principle any substance could be analysed into its elements, in practice the theory was of limited utility. When chemists decomposed a substance, the typical result was several products, none of which was a pure element; moreover, the products often varied depending upon the method of analysis. Chemists required a theoretical framework more directly related to reagents found on laboratory shelves. One such scheme was provided in the table of chemical affinities systematised by the French chemist Etienne-François Geoffroy (1672–1731) around 1718. The principle was a simple one: affinity tables were divided into several columns, each headed by a common reagent a mineral acid, an alkali or a metal. Below each reagent were listed the principal substances that would react with it, in order of decreasing affinity. The tables encapsulated information about chemical reactions and facilitated prediction by analogy, but even more importantly, their use encouraged chemists to distinguish substances by systematic reaction with a series of reagents, rather than mere observation of their individual properties.

The brightest prospects for chemistry, in Venel's eyes, lay in the teachings of Johann Becher (1635–82) and Georg Stahl (*c.* 1660–

1734), founders of the doctrine of essential chemical principles. In the eighteenth century, substances were classified into generic groups defined by a shared property, that is, according to whether they were earthy, metallic, combustible, acidic, caustic, and so forth. In essence the doctrine of Becher and Stahl embodied the defining properties in a set of hypothetical, quality-bearing material 'principles'. The notion had long roots reaching back to the Paracelsan *tria prima* and beyond to Aristotelian essences, but Stahl had cast the doctrine into a viable form by his dramatic success with the inflammable principle. He did not invent the concept, though he gave it a new name-*phlogiston*. Stahl's celebrated achievement was his taming of the elusive principle. Phlogiston could not be isolated or displayed, but Stahl showed that it was identical in the three realms of nature and could be passed from one combination to another at the chemist's whim. Phlogiston from charcoal could restore the principle that a metal lost upon calcination or could regenerate sulphur from its combustion product, sulphuric acid. It was the chemist's ability to manipulate phlogiston that made its existence plausible.

Lavoisier entered chemistry during the 1760s at a time when the efforts of the encyclopedists promoted a growing appreciation of its utility and philosophical interest. He did not arrive by the traditional routes of medicine, pharmacy or metallurgy. Instead he received a broad exposure to mathematics and the natural sciences through formal education at the College des Quatre Nations and the family connections that introduced him to leading Parisian men of science. The deepest influence upon the young Lavoisier was probably that of Jean-Etienne Guettard (1715–86), who engaged him as an assistant on excursions with Rouelle, Lavoisier continued his education by reading extensively in mineralogy and chemistry throughout the middle years of the decade. There is a common impression that he was disenchanted from the start with current chemical concepts. Surviving documents do not support that contention: he defended the theory of the four elements and used the concepts of phlogiston and chemical affinity in the same routine way as his contemporaries. However, his approach to the chemical literature was exceptional, for he read with an unusually critical eye, alert to discrepancies among different authors or to 'singular' phenomena, that did not follow simply from accepted doctrine. Such loose ends appealed both to Lavoisier's intellect and to his ambition as opportunities for 'a beautiful course of experiments'. What further set him apart from typical students of chemistry was the taste he had developed for the rigour of mathematical logic and the precision afforded by instruments in physical experiments. He found such rigour and precision were often lacking in chemical investigations, but they became the hallmark of his own work. Lavoisier's experiments did not yield new substances, such as the gases and organic acids isolated by his distinguished contemporaries Joseph Priestley (1733–1804) and Carl Wilhelm Scheele (1742–86). Throughout his career Lavoisier's efforts were focused upon *problems* and his discoveries were solutions to those problems.

One of the most consequential of the problems that captured Lavoisier's imagination during that youthful period was inspired by reading an essay on the elements by the German chemist, Johann Theodor Eller (1689–1760). Eller had revived van Helmont's attempt to reduce the number of *bona fide* elements, by claiming that water was transmutable into both earth and air. Lavoisier was intrigued by Eller's argument that air was 'factitious', that is, composed of water combined with fire which made it expansive. However, he rejected Eller's conclusion that air was the same thing as water vapour, suspecting instead that it might consist of an *unknown* substance rendered into vapour form by union with the matter of fire. This notion, conceived in 1766, was the seed that Lavoisier gradually articulated into his theory of the vapour state.

At that time Hales's discovery that air could be elicited from many substances had spawned quite different traditions in Britain and France. In Britain it inspired pneumatic chemistry: the isolation and study of different kinds of air. Joseph Black (1728–99) was the first to characterise a species of air differing from atmospheric air; it would not support combustion but had the property of precipitating limewater. Because it entered concrete substances like chalk or magnesia alba, he called it 'fixed air' (carbon dioxide). In 1766, Henry Cavendish (1731–1810) similarly described a species of air that was inflammable. But Joseph Priestley, in the early 1770s, emerged as the most creative discoverer and leading authority in pneumatic chemistry, by his isolation of a series of airs and his tests to distinguish them. Meanwhile, in France, Lavoisier remained unaware of the post-Halesian work until the end of 1772. Disseminated in Rouelle's lectures, Hales's results gave rise not to a tradition of pneumatic researches but to a theoretical dispute. Was the air obtained from bodies truly combined in them, or was it merely physically trapped in their pores? Then again, might the air be factitious, that is, not actually present in the bodies at all, but formed during their decomposition by the union of some other substances? Lavoisier's hunch of 1766 offered a simple resolution to the fixation of air. Removal of the combined fire that made air expansible would reduce it back to a concrete substance capable of entering chemical combination in the ordinary way. This promising notion did not immediately lead to an experimental investigation, perhaps because of competing demands on Lavoisier's time, but perhaps also because a line of attack had not yet crystallised in his mind.

Only in the summer of 1772 did Lavoisier undertake a sustained programme of research to elucidate the chemical role of air. The origin of his pivotal experiments on combustion and calcination, executed in the autumn of that year, has been the subject of controversy. H. Guerlac placed those experiments in a new light by showing that during the previous summer Lavoisier drafted a theoretical paper on the chemical role of air and fire; moreover he specifically intended to extract air from a variety of minerals by means of a large burning lens. It remains a moot point how Lavoisier proceeded from those plans to his crucial discovery (deposited at the Academy of Sciences in a sealed note on 1 November 1772) that combustibles or metals, when burned, gain weight by absorption of air. One view (favoured by Guerlac) is that the mysterious weight gain of calcined metals aroused Lavoisier's curiosity; alternatively, it has been proposed that an interest in combustion or in the formation of acids actually provided his starting point. Although none of these topics was foreign to Lavoisier as a possible area of study, the evidence does not support an explicit intent on his part to investigate any one of them. A more likely scenario is that continuing themes of his research were channeled in new directions by a series of contingencies, leading to an unforeseen discovery.

In the spring of 1772 Lavoisier 's name had been put forward to a committee set up by the Academy to investigate the puzzling disappearance of diamond when heated in a potter's kiln. Members of the committee requested that a large burning lens belonging to the Academy be brought out of storage in order to extend the study of the diamond and other gems to higher temperatures. Lavoisier seized the opportunity as a new means to pursue an old interest in minerals. Hales's results on a few minerals convinced him that many contained air – which he hoped to disengage with the burning lens and capture in a pneumatic device. He suspected, too, that metals (known to effervesce with acids) might also yield 'fixed air'. These were the most original of the experiments he proposed to try with the burning lens; but because of the difficulty of finding vessels to withstand its intense heat, they were never completed.

Lavoisier's participation in the burning lens investigations brought him into frequent contact with Pierre Mitouard, a Parisian pharmacist who had recently obtained a good sample of white phosphorous from Germany. Several authors had noted that phosphoric acid obtained by spontaneous combustion was heavier than the original phosphorus; some attributed the gain to absorption of air by the acid, others to attraction of atmospheric moisture. Mitouard probably mentioned these opinions, which were not, however, a focus of his own research. Lavoisier, at the time preoccupied with the chemical role of air, purchased an ounce of phosphorus from Mitouard to try his hand at resolving the question. Initially, the solution evaded Lavoisier, for he had to expose his phosphorus to the atmosphere in order to burn it; the resulting acid absorbed moisture by deliquescence as it formed, gradually dissolving itself. It remained unclear how much of the increased weight was due to moisture and how much, if any, might be due to air. Never one to waste his efforts, Lavoisier saw that repeated combustions could yield a concentrated sample of the acid from which he could derive its little-known salts. Over the following weeks he laboriously collected his sample and began a paper on the formation of the acid and its salts. But as he proceeded, a resolution to his original problem dawned upon him. He placed his acid solution in a flask, marked its level and weighed it; he then filled the flask to the mark with distilled water and weighed it again. The difference in weight gave the amount of phosphoric acid in solution, which turned out to be markedly greater than the weight of phosphorus consumed an increase, Lavoisier concluded, that could only be due to absorbed air.

The discovery so excited Lavoisier that he dropped his study of phosphoric acid salts and deposited his unfinished paper at the

Academy (on 20 October 1772) to retain priority. Having demonstrated the absorption of air by burning phosphorus, he wanted to see if the same might occur with sulphur. A positive result with sulphur then led him to speculate whether what was true of phosphorus and sulphur might apply to *any* combustible that gained weight by burning. In particular, he suspected that the weight acquired by metals during calcination (in Stahl's theory merely a slow combustion) might involve absorption of air. Within a week he confirmed his conjecture by the reverse process of reducing litharge (a calx of lead), from which he obtained a volume of air nearly a thousand times that of the calx.

The result with litharge completed a striking reversal of Lavoisier's expectations of the previous August. At that time he had hoped to extract air from metals by intense calcination; he now saw that on the contrary, under such circumstances they *absorb* air. His understanding of the relation of metal to calx (or sulphur to its acid) was inverted! Success in monitoring the entry of air into combination, and his resulting insights into the composition of familiar materials, inspired Lavoisier to predict a 'revolution' in chemistry. It is often assumed from hindsight that he meant the overthrow of phlogiston. However, the evidence does not support that claim, because he believed initially that air entered a combustible *as phlogiston left*. Subsequently, his doubts about the role of phlogiston grew until, in the early months of 1773, he came to the verge of total rejection of the concept. Therefore, the dismissal of phlogiston was a consequence rather than a cause of his revolutionary programme (and did not become final for several years).

The scope of Lavoisier's programme is more fully set out in a research memorandum he drafted on 20 February 1773. In the interval he had read other discoveries relating to 'fixed' air (including the British work) which he now saw as isolated links of a great chain one he personally hoped to join together. The key to his anticipated synthesis was to concentrate upon the processes which absorbed air; in that way he hoped to learn about the *origin* of air in substances and to follow its progress from one combination to another. As promising starting-points, he singled out the processes of vegetation, animal respiration, combustion, calcination and certain chemical combinations. From the start Lavoisier had a larger goal than a new combustion theory. His vision embraced a new system of chemistry in which atmospheric air, or some component of the air, entered by traceable chemical paths into operations of the animal, vegetable and mineral kingdoms.

From one perspective, the remainder of Lavoisier 's career can be viewed as the fulfillment of that programme. His early publications aimed at demonstrating the role of air in calcination; his emphasis shifted in the mid-1770s to combustion and the constitution of acids, then in the 1780s to fermentation, vegetation and respiration. This interpretation stresses the continuity and momentum inherent in his programme. From another point of view, however, Lavoisier's early conceptions were far from the system that he eventually published in 1789 as his *Traite Elementaire de Chimie*. To that point his departure from convention consisted in (1) an emphasis upon air and an embryonic theory of its fixation, (2) an inversion of familiar assumptions about composition, and (3) a recognition of the incompleteness of Stahl's combustion theory. On the other hand, he had not yet distinguished 'fixed' air from ordinary air, still regarded air and water as elements, had not definitively rejected phlogiston and he still comfortably worked within the framework of chemical principles. Indeed, Lavoisier 's programme of 1772–3 might be characterised as an attempt to establish air as a chemical principle, much as Stahl had established phlogiston. The further transformation of his views resulted partly from articulation of his programme and partly from its collision with discoveries and ideas advanced by his contemporaries.

Response to Lavoisier's innovations occurred in three phases. The first public disclosure of his work came in a paper on calcination read at the Academy's Easter public meeting in 1773. There he hinted at differences with Stahl and ventured that deeper study of fixed air would lead to a period of 'almost complete revolution'. That brash forecast drew an admonition from his academic colleague, Antoine Baume (1728–1804), who complained in print of certain physicists who wished to substitute fixed air for phlogiston. It may have been reproof from his colleagues, as much as the conceptual difficulty arising from the elimination of phlogiston, that led Lavoisier to avoid confrontation over Stahlian theory when he published an account of his research, *Opuscules Physiques et Chymiques* early in 1774.

Outside Paris, Lavoisier's early publications attracted attention primarily where his conclusions about calcination and combustion clashed with other views. The most prominent of the early encounters was with Joseph Priestley, whose concurrent investigations led to ideas on the role of air in combustion diametrically opposed to Lavoiser's. Like Lavoisier, Priestley was indebted to Hales for many of his experiments - which were conducted, however, from a distinctive point of view. Priestley set out not to reform chemical notions, but to explore the role of air in larger economies of nature. His experiments aimed at elucidating the constitution of atmospheric air, the nature of other kinds of air and their effects upon such processes as the respiration of mice and the growth of plants. Aware that certain processes, including combustion and respiration, 'spoilt' the air, he sought ways to restore its 'goodness', measuring the 'goodness' by diminution of a sample when mixed with a predetermined amount of nitrous air (nitric oxide). Priestley's work found enthusiastic emulators on the continent; Lavoisier himself first showed an interest in respiration after reading Priestley's early paper on gases and learned to distinguish gases qualitatively by repeating Priestley's experiments.

In his *Opuscules,* Lavoisier criticised Priestley's view that ordinary air was diminished during combustion by phlogistication and implied that the British philosopher had failed to note the *absorption* of a portion of air by the burning substance. Priestley was quick to deny any suggestion that phlogiston diminished the elasticity of air, pointing out that it had the effect of precipitating 'fixed' air from the atmosphere - hence the shrinkage. Differences between Priestley and Lavoisier were exacerbated in 1775 when Lavoisier published his famous paper showing that the red calx of mercury (upon reduction without charcoal) yielded not Black's 'fixed' air, but an air capable of supporting combustion. Priestley himself had obtained such an air from the red calx in August 1774 and had mentioned it in Lavoisier's presence during a Paris visit in October. In the second volume of his *Experiments and Observations on Different Kinds of Air* Priestley not only chided Lavoisier for failing to observe that the air was a species much better than ordinary air (Priestley proposed to call it dephlogisticated air), but also implied that Lavoisier ought to have acknowledged his own priority. Theoretical differences between Priestley and Lavoisier were therefore compounded by an apparent breach of 'philosophical' etiquette - plus perhaps, a rivalry for authority in the popular new field of pneumatic chemistry.

In its early manifestations, Lavoisier's programme was not perceived as a threat to the foundations of chemistry because its revolutionary implications were concealed and the validity of its limited claims disputed. A second phase in its reception was launched with his reading of a paper 'On Combustion in General' in the autumn of 1777. On that occasion Lavoisier made clear his intention to account for combustion without employing the concept of phlogiston; accompanying thermal effects were instead attributed to the matter of fire released as air relinquished the vapour state. For several years, Lavoisier refrained from direct confrontation with colleagues over the existence of phlogiston, electing instead to develop the positive implications of his theory. He proceeded with analyses of acids (showing that they contained dephlogisticated air, which he renamed *oxygen,* the acid-former), and elaborated his theory of heat in collaboration with Laplace. Around 1781–2 he sketched an outline of a systematic presentation of his views, in which for the first time he articulated his new understanding of the elements, defined pragmatically as simple substances that were the end result of chemical analysis. His list included the metals, sulphur and phosphorus as well as the more traditional earths and water. Despite the growing power and coherence of his system, he was confronted by rival theories and disconcerting counter-evidence. His colleague, Pierre Macquer, had advanced a compromise theory that granted the absorption of air in combustion (and hence could explain weight relations precisely as Lavoisier did), but retained phlogiston as the pure matter of sunlight (allowing him to account for reduction without charcoal). The theory attracted a following in France as a way of incorporating Lavoisier's compelling results within a conventional theoretical framework. Another serious competitor was the theory-favoured by Jean-Claude de Lametherie (1743–1817) in France and R. Kirwan (*c.* 1733–1812) in England – that inflammable air was pure phlogiston. Metals were known to release inflammable air when they dissolved in acid; even more compelling was Priestley's observation in 1782 that a calx of lead was successfully reduced by heating it in inflammable air (which was consumed as the metal formed). How could such phenomena be explained without the assumption of an inflammable principle in the metal? The second phase closed with a startling discovery that turned these phenomena to Lavoisier's advantage. Repetition of Cavendish's ignition of a mixture of inflammable and dephlogisticated airs, yielding water as a product, led Lavoisier to recognise that water was not an element but a compound of the two gases. So inflammable air evolved 'from metals' could be attributed instead to decomposition of water. This insight was supported by an analysis of water, by passing steam over heated iron filings. The water was decomposed, its oxygen combined with the iron and its inflammable air (which Lavoisier now renamed *hydrogen,* the generator of water) was set free in considerable quantity.

A third phase in the reception of Lavoisier's views began with public response to the alleged composition of water and to his provocative 'Reflections on Phlogiston' read to the Academy in the summer of 1785. In it he argued that Stahl's theory was based upon circular reasoning, that variations of phlogiston theory were incompatible with Stahl's and with one another, and that the principal phenomena of chemistry could be simply explained without phlogiston – a strong indication of its non-existence. Although Stahl's theory might have benefited chemistry initially, its retention had become an obstacle to further progress. The 'Reflections' produced an uproar of protest, at first within the Academy then later more widely with the paper's publication in 1786. The tacit truce on fundamental issues was abandoned. The existence of phlogiston was perceived as the central issue and the analysis of water as a crucial experiment. Stahl's followers objected that the formation of water from the two gases indicated that it was a constituent of the gases, rather than the other way around; in the alleged analysis of water, the abundant inflammable air came, they said, from the iron filings, not the water. The experiments were difficult to repeat and both the results and the interpretation remained contentious for several years.

As the deeper implications of Lavoisier 's programme spread and especially as a few prominent members of the French community declared their support, its threat could no longer be ignored. A schism began to develop within the French community. A couple of the Academy's younger chemists, Claude Louis Berthollet (1748–1822) and Antoine François de Fourcroy (1755–1809) – and many of its mathematicians and physicists – rallied to Lavoisier's side; from Montpellier Jean Antoine Chaptal (1756–1832) declared his support. Promotion of the new chemistry increasingly took on the aspect of a collaborative campaign; its advocates were dubbed 'antiphlogistians' and their activities were described in political metaphor. Active opposition to the new chemistry grew in proportion, finding its voice in the *Journal de physique,* under de Lametherie's editorship. Criticism was initially reasonable, stressing counter-evidence and theoretical objections, but as the strength of the antiphlogistic party grew, it sometimes turned acrimonious and personal.

Early in 1787, Louis Bernard Guyton de Morveau (1737–1816) travelled to Paris to consult leading chemists on the reform of chemical nomenclature that he had been urging for several years. Long a respected defender of phlogiston, Guyton had gradually modified his position, in the light of new evidence, until little separated it from Lavoisier's. In Paris he was finally won over. In collaboration with the antiphlogistians, he then drew up a new method of chemical nomenclature grounded on Lavoisian principles. The language reflected the new view of chemical composition. It was founded upon a set of operationally defined 'simple' substances', each given a simple name. Compounds received binomial names revealing their elemental constitution and assigning them a genus and species. So a lead calx, for example, was allotted to the genus oxide, species lead. It is the familiar naming system for simple inorganic substances such as sulfides, nitrites, nitrates and so forth, still in use today. Promoters of the new nomenclature saw it as a significant advance. Not only would chemistry become easier for beginners to learn, but the systematic and analytical nature of the language would also improve reasoning in chemistry. To opponents, on the other hand, the new terms had a barbarous ring. The system was objectionable because, from a practical point of view, experienced chemists would not understand the foreign terms; philosophically, the whole system was founded upon hypotheses which might collapse; politically, it was seen as a blatant attempt to impose the new theory by forcing beginners to learn the new language.

Despite vocal opposition, the trickle of converts to the new chemistry turned into a stream during the late 1780s. Its early promoters included several of France's most successful chemistry teachers: each year hundreds of students assimilated the new view through the able lectures of Fourcroy in Paris, de Morveau in Dijon and Chaptal in Montpellier. However, the transition was not simply a matter of an older generation dying off to be replaced by a younger and more flexible one. The community as a whole shifted to the new system, though not all at once, nor with uniform enthusiasm. Conversion proved more difficult for older chemists, who were possessed of deeper attachment to the old theory and acute awareness of the limitations of the new. Yet many of them managed partial adoption of the antiphlogistic system as its experimental claims were confirmed and its theory clarified. The campaign was significantly advanced by publication of several antiphlogistic treatises especially Lavoisier's *Traite* providing a coherent overview of the new system to combat the fragmentary and often distorted perceptions of it in circulation.

The struggle for the new chemistry in France was effectively accomplished by 1789–90, although some prominent figures persisted in their opposition. During the early 1790s many sceptics quietly began to employ the new concepts. By the middle of the decade, even that inveterate opponent, de Lametherie, found himself obliged to use the new nomenclature in order to be understood by readers of the *Journal de physique*. Elsewhere, dissemination of the new theory proceeded at different rates, according to the commitments of local communities and the quality of their communications with France. In Britain, where mutual interests and excellent lines of communication facilitated awareness, the course of debate closely paralleled that in France; antiphlogistic publications were rapidly translated into English. By the early 1790s the new chemistry was well established in Edinburgh and Glasgow where it was endorsed by Black and his students. Prominent English philosophers were reluctant to concede the superiority of the Lavoisian system: Cavendish appears to have abandoned the defence of phlogiston (without championing oxygen) by about 1787; James Watt adopted the new ideas with reservations, while Richard Kirwan capitulated to the antiphlogistians in 1791 but never became a whole-hearted supporter. Of course, Priestley and his friend James Keir remained against it to the very end. The stubborn German opposition to the antiphlogistic chemistry is well known. Ironically, it owed much to the self-sufficiency of the German chemical community (reinforced by emergent cultural nationalism), which did not encourage chemists to pay attention to foreign developments. More generally, the implications of the Lavoisian system tended to be ignored until they were championed locally by a few individuals. Typically, those individuals were young; many had travelled and come into contact with antiphlogistians. Once the issue had been raised, arguments followed similar patterns to those in France and Britain, but with local variations; the existence of phlogiston always remained at the heart of the debate. By the close of the century the new chemistry had penetrated even the remote corners of Europe and America. Paris emerged as the leading European centre for chemical instruction; students flocked there to hear lecturers by Lavoisier's supporters who, since the political revolution, had dominated all of the Parisian scientific institutions.

The new chemistry was established amidst dissension; today its interpretation continues in controversy. There is consensus that the overthrow of phlogiston is too narrow a focus, but no suitable synthesis has taken its place. Novel (and sometimes incompatible) interpretations have been proposed. The Lavoisian achievement has been portrayed as the reaffirmation of material principles over British attempts at mechanisation, or alternatively, as the subordination of chemistry to physics. Several authors shift the emphasis away from phlogiston or combustion to the nature of the vapour state, or the theory of acidity, or the new view of composition. But insistence upon any one of these aspects would be just as restrictive.

A more promising approach would be to treat the transformation as a complex process with several layers of meaning perceived differently according to time and place. Lavoisier first spoke of 'revolution' when he saw a way to elucidate the nature of air; ultimately he succeeded in his goal of establishing air (or rather oxygen) as a key chemical entity. His path led through detailed studies of calcination, combustion, acids, fermentation, organic analysis and respiration; it was his specific claims about these processes that constituted the manifest implications of his work. But his programme had deeper consequences (some unforeseen in the autumn of 1772), notably the elimination of phlogiston and the overturning of the traditional hierarchy of composition. Close colleagues had an early glimmer of these implications, but the community as a whole remained unconcerned until the direct attacks upon phlogiston and the publication of the nomenclature. On a disciplinary level, Lavoisier employed physical methods, received support from physicists and brought chemistry much closer to its sister science an intrusion that was resisted. Even more imperceptibly, Lavoisier 's synthesis undermined the entire doctrine of chemical principles, substituting a combinatorial view in which the properties of simple substances bore no necessary relation to those of their compounds. These changes were considerable. Nineteenth-century historians, comparing the concepts, language and methods of Lavoisier's chemistry with its predecessors, saw in them the beginnings of modern chemistry.

The Chemical Revolution is a classic instance of conceptual change in science – one of the first to be foretold. Venel had called for a breakthrough that would exploit the distinctive methods and concepts of chemistry to establish it as the independent peer of physics. Ironically, Lavoisier found that breakthrough in the chemical role of air, but pursued it by the methods of experimental physics. His innovations transformed the structure and language of chemistry, generating a crisis that split the community. Chemistry emerged from the conflict as a more mature discipline with the public recognition Venel had desired. A new historiographic synthesis is called for, one that neither imposes another strait-jacket nor trivialises the episode, but incorporates the full richness of the transformation.

12

Cell Theory and its Development

Introduction

Cell theory has passed through various stages of understanding during the three centuries since researchers first noticed the existence of cells. Likewise, the role of cells in development has also gone through different interpretations. Theories have included an elementary conception of basic microscopic units in the mid-seventeenth century, a more fully articulated cell theory in the early nineteenth century which held that cells are the basic building blocks of living organisms, a later nineteenth-century conviction that the actions of cellular material actually bring about organic development and differentiation, and the idea that cells hold a key to evolutionary development as well. Controversies and disagreements about detail have occurred, but some version of cell theory and the fundamental role of cells in development has persisted.

This essay will examine the intersection of studies of cells and studies of development, which necessarily leaves out much of both cytology and embryology. Yet the study of cells and their role in individual development has raised important questions and has shaped both the direction of cytology and the study of embryology. It is worth focusing on the intersection and on those few but valuable historical examinations of both cell theory and development.

Cytology and Cellular Stuff

In 1665, Robert Hooke's *Micrographia* (1665) first brought cells to public attention. Intended as a popularly accessible book rather than a specialised report, the *Micrographia* illustrated and described cells which Hooke had studied with his microscope. Taking a thin slice of cork, for example, he observed what looked like a honeycomb of pores, which he called cells since they were essentially spaces surrounded by walls and not unlike the monastic cells of the time. Then he observed similar textures in a number of other plant forms as well. Though convinced that these cells must have a proper purpose, this English observer was not quite certain what that purpose might be. He concluded that they might serve as channels to carry fluids through the plant material in just the way that arteries and veins move fluids through the animal body. Hooke thus established that structural units called cells exist in some organisms, but was not confident about further conclusions.

A few years later, other observers turned their microscopes to organic material as well. In particular, the Englishman Nehemiah Grew and the Italian Marcello Malpighi both began to observe more clearly the detailed structure of plant cells with their work which they called vegetable anatomy. Because the cellular structure of plant material is more obvious than that of most animals, they each clearly observed cells. *Malpighi's Anatomes plantarum* (London, 1675, and 1679) referred to the 'utricles' (or cells) and to the 'basic utrical structure' of the plants. At the same time, in work also communicated to the Royal Society of London, Nehemiah Grew addressed the question raised by Hooke as to whether plants have circulatory systems like animals. He did not find any valves or perfectly analogous vessels to answer that question, but he did provide more detailed and definitive descriptions of the cells in his work which culminated in *The Anatomy of Plants* (London, 1682). The contemporaneous work by Malpighi and Grew set the standard for discussion of cells as structural units for some time. It also demonstrates the close connection between microscopic exploration and cell study in the early years.

Since microscopic aid is valuable for seeing most cells, the increasing interest in microscopy in the seventeenth and eighteenth centuries brought with it an enthusiasm for describing the structure of materials, including the apparent cellular structure of living materials. Thus, some historians have cited the crucial directive role of technical advances, while others have argued that ideas preceded technical innovation. Whatever the technical contribution, questions arose among the microscopists. While Hooke, Grew and Malpighi located cells in plants, others questioned what those cells were: whether the basic units of life or parts of an interconnected fabric; whether accidental and occasional or ubiquitous structures; or whether perhaps only unimportant microscopic artifacts. Others asked whether cells exist in animals as well. The late eighteenth century brought discussion of 'globules', for example, as alternative fundamental material structures. Others saw 'fibrilles' as the more proper units of organic material, and some invested the latter with vital properties such as 'irritability'.

This proliferation of ideas creates problems for the historian who wants to examine cell theory, because it remains unclear what to count as cells. Do we include only those discussions that actually label their entities as 'cells', or do we also include references to 'globules' or 'fibrilles'? When examining cell theory *and* development, such work as Caspar Friedrich Wolff's seems important since he regarded embryos as made up from globules, which produced a sort of 'cellular tissue' analogous to plant cells. And yet the globules do not really correspond to visible cells or to basic organic structural units, as Shirley Roe has clarified in *Matter, Life, and Generation*. It is reasonable, therefore, to exclude Wolff as belonging to a different tradition, with different concerns and commitments, and to remain with the microscopic-based study of what cells are and what their role is in development.

In order to begin addressing those developmental questions and to turn those structural spaces surrounded by walls into something more functional, it was useful, firstly, to fill them with something and, secondly, to determine how they arise. Felix Dujardin began to answer the first question in 1835. His 'sarcode' was the material which fills living cells: 'a glutinous, diaphanous substance, insoluble in water, contracting into spherical masses, sticking to the dissecting needles and letting itself be drawn out like slime, and, finally, being found in all the lower animals interposed between the other structural elements'. Dujardin's work on infusoria, designed to examine the internal structure of that organic group, also raised the question of whether similar substances exist in other organisms. In 1839, the Bohemian researcher Jan Purkinje confirmed this and identified the fluid stuff inside the cell walls as 'protoplasm', a concept which Hugo von Mohl extended and developed into its modern form.

Both Dujardin and Purkinje suggested that the internal fluid protoplasm might be the basic material of life, but it remained for the German cytologist Max Schultze to carry that suggestion into a fully developed theory – the protoplasm theory. In a paper of 1861 directed at clarifying the nature of a cell, Schultze acknowledged that a cell had been defined as a structure with 'membrane, nucleus, and contents', but that in fact only the nucleus and protoplasm appeared as universally basic. Thus 'A cell is a lump of protoplasm inside of which lies a nucleus'. The membrane or cell wall separates the cell contents from the external environment but is not, in fact, really necessary since the unique protoplasmic substance is kept distinct from the surrounding material simply by the fact that it does not mix with water. He pointed out, in particular, that cells which are undergoing division have no membranes. For Schultze, the crucial internal contents do not consist of a simple watery fluid but rather of a thick viscous mucous substance, comparable to soft wax. This protoplasmic material is basic to life, while the nucleus plays an as yet unknown role. Though he worked with a variety of animals himself, Schultze also cited work on plants to demonstrate the uniform protoplasmic basis of all living material in protoplasm.

The cell had acquired a very different consistency and role ascribed to it from the one Hooke had suggested nearly two centuries earlier. Thomas Henry Huxley's popular essay of 1868 suggesting that protoplasm was the 'physical basis of life' brought the protoplasmic view to wide attention stimulated further discussion. What was the relative importance of nucleus and protoplasm, and in what sense were the cells the *units* of life if protoplasm was the *stuff* of life? The different lines of research, with respective emphases on plants or animals, or on adult or dividing embryonic cells, for example, gave rise to different views of the cells and their importance. Generally, those stressing the central importance of protoplasm de-emphasised the significance of cellular units and of what has historically been called 'the cell theory' which an alternative tradition stressed. It is the latter tradition which provided the strongest basis for progress in interpreting the cellular role in development and which must remain central for historians concerned with both cells and development.

Cell Theory and Cellular Development

At the same time as the theory of cell substance was evolving, in the mid-nineteenth century, a very different theory of how cells come into existence was also emerging. The German botanist Jakob Mathias Schleiden and the zoologist Theodor Schwann collaborated to present what has come to be labelled as the cell theory. Their theory depended very directly on their definition of how cells come into existence as part of their life cycle and it differed from the protoplasm theory in emphasising the essential role of the cell wall for defining this basic unit of life.

Schleiden was one of those young biologists of the early nineteenth century who could not accept the then traditional preformationist interpretation of the origin of organic form. Preformationists of the previous century had answered the age-old question: 'how does an individual come to have his particular differentiated form?' with the answer: 'from pre-existing form'. An individual organism inherits its form from its parent. Logically, therefore, an individual cell would also inherit its form from earlier cells. But Schleiden and others, such as the German embryologist Karl Ernst von Baer, had begun to reject such a preformationist interpretation and to turn instead to epigenesis. An epigenetic position holds that form emerges anew, shaped by the interaction of living internal material and its external environment. As an epigenesist, then, Schleiden wanted cells to emerge anew in each generation. He maintained that those newly-arising material cells serve as the fundamental units of both organic structure and organic function for all of organic nature. Cells are the basic units of life. Schleiden and his collaborator Schwann further insisted that all cells in both plants and animals originate according to the same general set of procedures; all cells must therefore be fundamentally the same sorts of things. Unity of nature reigned, as it did for most early-nineteenth-century German naturalists.

Schleiden's theory of free cell formation held that the following process occurs for plants: inside the contents of a cell a granular substance arises; by accumulating surrounding material, this gives rise to a nucleus; as the nucleus grows a cell forms and a surrounding membrane appears to set it off as a new cell. Schleiden maintained that this process occurs as an 'altogether absolute law'. The resulting cell consists almost incidentally of those internal contents labelled by others as protoplasm, but also necessarily includes a nucleus and cell wall.

Stimulated by his discussions in Berlin with Schleiden, Schwann further developed the 'Schleiden-Schwann cell theory' with his animal investigations. He suggested that, beginning with a structure less substance or 'cytoblastema', a dark granule arises, which in turn gives rise to a nucleus. Then layers of substance accumulate around this core to produce a full cell. For Schwann, this process takes place in material surrounding, rather than inside, the old cells, so that in this respect, he differed from Schleiden. For both, nonetheless, the new cells are really new and not simply inherited or pre-existent in any sense. They both maintained that cell formation strictly follows the inorganic and hence materialistic process of crystallisation. Then once formed, the cells serve as structural and functional units for living organisms.

Published together and translated in 1847, the claims of Schleiden and Schwann became the basis for a 'cell theory', which held cells to be the fundamental organic units common to all living beings and as developed by 'free formation' out of formless cytoblastemic substance in the same basic way. Some challenged the exact definition of the cell – whether it requires an enclosing membrane or wall to define it, for example. Others questioned the mode of cell origin whether always *de novo* rather than from pre-existing cellular material in particular. Work on plants and animals also diverged as researchers questioned the respective natures of plant and animal tissues. Yet most researchers had nonetheless begun to accept some version of cell theory by the mid-nineteenth century. The problem remained of sorting out the details and of elaborating the extent to which cells function as *the* fundamental units of life and living processes. In particular, Jacob Henle and Albert von Kolliker adapted the theory into classic full discussion within anatomy and histology, but cell theory as a basis for development took longer to elaborate. Maintaining a focus on cells, the German zoologist Robert Remak and the pathologist Rudolf

Virchow moved towards a different and eventually more widely acceptable interpretation of cell theory, which proved more useful for embryology.

Remak published the results of his extended study of freshly laid frogs' eggs and concluded that the cells in his embryos developed in a different manner from the one that Schleiden had described for adult plant cells. In citing the substantiating support of other researchers' results as well as his own earlier work, Remak endorsed an endogenous theory of cell formation, whereby cells form only from material internal to other cells. Indeed, all cellular development begins with the fertilisation of the egg cell, while subsequent cell cleavages result from a series of divisions directed by the nucleus (*Untersuchungen uber die Entwicklung der Wirbelthiere* (1850–55)).

Also in an essay of 1855 and with much more detail in his classic work *Die Cellularpathologie* (1858), Virchow agreed with Remak and other cytologists who had begun to question free formation, asserting that all cells arise only from other cells: 'Omnis cellula a cellula'. Cells do not crystallise in any intercellular cytoblastema, and since cells require both a nucleus and membrane, they must begin as more than tiny material nucleus-producing granules. Life is continuous as one cell gives rise to another, one generation to another. The complex and responsive cellular material is not only the basic *structural* unit of living material, it can also be the basic unit of *life.*

Virchow and others had come to the conclusion that cells somehow divide. Rather than new cells crystallising in a strictly materialistic manner around a granular core, Virchow saw the existing cellular material accumulate new material and grow larger, eventually reaching a point at which it divides. According to this vitalistic interpretation, living material does not emerge from non-living material. In fact, the history of the cell theory in the nineteenth century became closely tied with arguments about materialism vs. vitalism. According to Ackerknecht 's interpretation, Virchow 's particular interpretation of cell theory reflected his political vision of the role of individuals within the state.

Not everyone who accepted the doctrine that cells arise only from other cells also agreed with Virchow's politics or his form of vitalism. But continuity of cellular, living material made the epigenesist's task easier. If cells provide continuity from one generation to the next, then the epigenesist need not explain the production of form from completely homogeneous matter. Instead, he may assume the existence from the beginning of something which is already living, inherited from the past. As John Farley has pointed out, 'the re-emergence of sex' and sexual reproduction as a driving biological problem in the 1870s, accompanied by improvements in microscopic and cytological techniques and hardware, served to refocus attention on the cells notably on the egg and sperm or ovule and pollen cells – and their roles in development.

What resulted, as researchers embraced Virchow 's view of cell formation and substantiated it with observations resulting from improved microscopic techniques, was a move towards emphasis on cell division. As it stood at mid-century, the cell theory depended on an interpretation of how cells develop, but did not provide significant clarification of how individual organisms develop and become differentiated. If each cell grows by accretion of material, then the question remains of how cells mutate, and in such a way that a whole organism is created. What guides development?

Pursuing epigenetic interpretations of development, researchers began to accept that cell division may be the key to development of individual organisms and hence the key to what differentiates life from non-life, namely, the ability to reproduce successive generations of individuals. By the 1870s, enthusiasm for the materialistic Schleiden/ Schwann interpretation of cellular development and for protoplasm ideas had faded. Research had begun to enter a new stage, exploring the importance of cell division for development. This change in problems and approaches, with attendant shifts in meanings of words and in emphases, creates pitfalls for unwary historians who cast backwards for the roots of today's ideas. Unfortunately, biologists who have written casually about the history of cell theory have also fallen into that trap and have made the picture seem rather clearer than it was. It is important to recall that 'cells' do not even refer to the same thing as we move from the seventeenth to the nineteenth century, for example.

Out of the proliferation of studies in the 1870s, two very different lines of research emerged which need to be examined as separate traditions. One, growing out of the cytological tradition, carefully examined the structure of the cell. Using histological techniques for fixing, staining, preserving and cutting cells, the cytologists began to be able to look more deeply and to discern fine differences between cells. The nucleus attracted the first and most concentrated attention, but gradually other cellular parts also attracted interest. And with their discovery, researchers began to ask also what the cellular parts were for.

A second line of research took a wider look. Instead of focusing down into smaller and smaller parts, this group asked how the cells fit within the whole organism. Of particular interest for our purposes, embryologists began to see that distinctions between different cells might begin to explain established facts of heredity and evolution, since an individual organism begins as a single egg cell which is the product of an evolutionary past and of inheritance from its parent. That individual egg cell is also the beginning of a new individual and hence must contain whatever is needed to pattern that individual. This second group of researchers began to examine more closely what happens in the course of cell division: what, in fact, happens as the one original egg cell divides? Different groups focused respectively on the nucleus or on the whole cell.

Cell Theory and Development into the Nucleus

The cell nucleus seems first to have attracted the serious attention of British botanist Robert Brown. In 1831 he determined that the nucleus is, in fact, an important part of at least living cells and not just an artifact of microscopic observation. Brown was the first to label this body the nucleus and to observe systematically that it occurs in a range of types of organisms. Yet he did not develop his observations into a more general theory, nor did he speculate about the significance of that nucleus. In time, others did, as they also extended his sketchy suggestions concerning fertilisation as a result of kinetic interaction of ovule and pollen in plants.

In fact, work on fertilisation highlighted the first line of research into cell theory and development. Advances in microscopic techniques such as improved sectioning, fixing, and staining as well as the advent of the oil immersion lens, allowed a closer look at cell contents. Observers could now distinguish more than just a general nucleus, protoplasm and membrane. Thus, the 1870s and 1880s brought a flurry of new studies, of which those by the Belgian Edouard van Beneden, the Germans Walther Flemming and Oscar

Hertwig, the Polish/German Eduard Strasburger and the Frenchman Hermann Fol were particularly important. Their technical disagreements, aggravated by imperfect microscopic evidence, make it more difficult – and also particularly valuable for historians to clarify the threads of thought during this time.

Each of these men declared that the egg and sperm cell for by then fertilisation was generally agreed to involve the union of those two cells contain a distinct nucleus. Therefore, the nucleus does not come into being simply for the purposes of cell division and then disappear, as some believed, but maintains its individuality across generations. And yet, Fol, at least, denied that the structure of the nucleus has any purity or continuity and that the nucleus could play any significant role in heredity. However, as he continued work through the 1880s, Fol began to see the centrosomes as permanent structures, with each parent cell contributing two centrosomes or one which divided soon after fertilisation. Centrosomes might play a role in heredity. Hertwig disagreed, maintaining that the nucleus itself and its components have continuity but that centrosomes do not. Furthermore, for both Hertwig and Strasburger, the nucleus served as the bearer of heredity while for Fol it did not. When Hertwig, Strasburger and others actually saw the nucleus divide during cell division, that evidence considerably strengthened their case for a nuclear role in heredity and subsequent development.

If the cell, and particularly the nuclear part of the cell, really exists as an identifiable and continuous entity, and if fertilisation actually involves the joining of a nucleus from each of two parents, then the nuclei must play a basic role in heredity. But how? And how does the stability between generations, assumed to be brought by heredity then translate itself into the differentiation of individual development? For this group of researchers, inquiry began to focus on details of the process of nuclear division.

The occurrence of mitotic nuclear division (or mitosis) had been recognised as early as 1873. Many researchers immediately attacked the subject and determined that nuclear division is basic to cell division. Strasburger insisted that nuclear division occurs transversely, that is, across the chromosome, dividing it into two separate pieces of different materials. In contrast, van Beneden and Flemming insisted on longitudinal division. Van Beneden observed the movement of chromosomes as they moved during cell division. Then Flemming provided a striking set of studies demonstrating the stages of mitotic division, which he presented in three papers published from 1879 to 1881. He believed that this indirect division of what he called chromatin (the stainable nuclear material, which Wilhelm Waldeyer labelled as chromosomes in 1888) provided the basis for the process of cell division in all forms of cells. By the 1880s a group of researchers had decided that the nucleus plays a role, if not *the* critical role, in cell division generally and thus in the development of individuals. Flemming's *Zellsubstanz, Kern und Zelltheilung* (1882) provides an excellent review of work prior to that time.

Furthermore, by the late 1880s it began to seem as though the nucleus also plays a or the - critical role in moving from one generation to the next. Van Beneden and Theodor Boveri from 1887–8 and Hertwig in 1890 had each clearly observed that something unique happens during the production of germ cells. During this process of maturation division when one cell gives rise to the ripe germ cells, each chromosone divides once while each cell divides twice. Thus, each ripe germ has only half the original full complement of chromosomes. The chromosomes must play some important role in heredity to result in this complex process, but what? Earlier, equally careful cytologists such as Strasburger had insisted on transverse division of chromosomes and did not see an important difference between germ cell and adult cell production (or mitosis and what was later called meiosis). Only gradually during the 1880s and 1890s did researchers work out details of the structural and chemical constitution of the chromosomes and of the nucleus. And only after 1900 did theories begin to emerge, explaining the chromosomal role in heredity. The work of the British researchers John Bretland Farmer and John E. S. Moore first explicitly stated what happens in reduction division and the workings of the 'maiotic phase'. Their classic paper 'On the Maiotic Phase (Reduction Division) in Animals and Plants' also provides a useful survey of material to date.

Yet such detailed researches into the nucleus and nuclear division focused more and more narrowly on the fine structure and parts of the cell. This concentration drew attention away from development, partly because no one saw how nuclear change translated into epigenetic developmental change. By the 1890s, it was the second line of research, looking at whole cells, that still concentrated on cell theory and on development.

Cells's Role in Development

Cell theory, as American cytologist Edmund Beecher Wilson wrote in his classic *The Cell in Development and Inheritance,* was the second great generalisation of biology in addition to evolution theory. He realised that only recently had the two ideas begun to converge in a way that promised major progress for biology. In addition to those researchers concentrating on applying new microscopic techniques to observe more clearly the fine structure of intracellular material, others such as Wilson focused on the cell as a whole or on the organism as a whole. A tradition emerged, directed at showing how the different cells resulting from division of the egg give rise to different parts of the body. The researchers participating in this tradition were concerned with embryology (and evolution) rather than with the earlier stage of fertilisation or with nuclear change. They included especially a group of Germans and a group of Americans.

Embryologists had concentrated on the collections of cells making up germ layers rather than on particular cells. For one thing, it was simply too hard to see individual cells and what they did during development. Furthermore, evolutionary theory and Ernst Haeckel's particular interpretations of what happens suggested that the broader-reaching germ layers rather than individual localised cells held significance for understanding evolution and genealogical relationships. How could the peculiarities of one individual cell be the product of evolution, selectionists would have wondered?

Stimulated by improved research techniques, embryologists began to ask what happens to individual cells. Given that the egg and sperm and their union are all cells, perhaps careful detailing of what happens to cells in development could be worthwhile. A group of German physiologists and zoologists began to study *Entwickelungsmechanik* or *Entwickelungsphysiologie.* Alongside the work of researchers including Wilhelm His, Eduard Pfluger, Gustav Born and Hans Driesch, Wilhelm Roux devised an elaborate theory of the action of blastomeres, cleavage and then of cells and cellular parts and their roles in development.

Roux and other developmental researchers studied the factors which cause the egg cell to divide and how this division occurs. They investigated the effects on development of conditions such as experimentally-altered gravitational field, artificial light source, altered nutrients and chemical manipulations. With manipulative experimental studies, the researchers could begin to establish that not only does the individual organism respond to external conditions but it also exhibits a great deal of internal control over its development and differentiation. Eventually, Roux constructed a theory of mosaic development, according to which each cell division separates off cellular material which then, because of its constituent material, develops in a unique and appropriate way.

In particular, Roux performed a famous experiment in which he used a hot needle to kill one of the two cells resulting from the first cell division of a frog's egg, as discussed in his summary work *Die Entwickelungsmechanik der Organismen*. He left the dead material but it did not grow, so he considered that the cell had been functionally eliminated. The other cell developed into precisely the half embryo that it would have done under perfectly normal conditions. Therefore, Roux concluded, it must have been something inside the cell itself, part of its own material which directed it, in this case, to develop as a half embryo. He presumed that the same constituent would also direct a normal cell product to develop in its appropriate manner. Thus, cells are parts of organisms but develop largely independently, according to their own internal instructions. Those instructions, he decided, came from the nucleus and specifically from the chromosomes.

For Roux, the chromosome is a complex mixture of different chromatin granules, which represent different qualities. These align themselves in preparation for cell division and then divide. Division may occur quantitatively, so that the original granules reproduce and then separate, with all qualities represented in each of the two daughter cells. Division may also occur qualitatively, so that the two daughter cells end up with sub-sets of the original materials and are thereby differentiated. The latter process occurs during embryonic development. By this process, each nucleus receives its own unique set of chromatin, and the course of development involves a successive separating-out of original pieces into the various cells. Because it contains its own chromatin qualities, each cell is capable of self-differentiation and experiences some autonomy even while it is part of the whole, complex organism.

In *Das Keimplasm*, August Weismann developed a similar theory, carrying much further the idea that separate bits of chromatin represent different qualities. He constructed an elaborate hierarchy of cellular parts with biophores as the basic units. These are then aggregated progressively into determinants, ids, and idants, the latter of which finally represent the visible chromosomes. The complex chromosome structure is inherited from one generation to the next because the cells of the germ plasm retain the full complement of sub-units, while body cells receive only part and become differentiated accordingly. Mitosis serves to distribute the smallest differentiated bits of chromatin to the different cells and thus becomes the mechanism for effecting individual development as well as intergenerational heredity. Eventually, each cell will have only one kind of determinant in it, which will give it its specific character.

For both Roux and Weismann, whose ideas came to be labelled the Roux-Weismann theory in the mid-1890s, the cell remained basic. Chromatin carries the material of heredity, but it is the division into separate cells that brings development. Cell theory was once again a basic assumption for these biologists, but with a very different emphasis. Roux and Weismann had explained a mechanism by which individual cells become differentiated and hence by which complex organisms grow, but many others did not find this explanation satisfactory. The resulting discussions provide valuable material for historians examining cells and development.

Hertwig and others in Germany rejected the Roux-Weismann interpretation of development, saying that it really explained nothing. For Hertwig, Weismannism represented a hopeless sort of preformationism which throws everything back on to the nucleus and chromatin rather than on to preformed little beings, but which is preformationism nonetheless. Questions of cells and development therefore became tangled with traditional debates about whether epigenesis or preformation better explains development. A further question arose as to whether perhaps cell theory itself had been overtaken. Perhaps the cell is not the smallest basic unit of life after all. Roux and Weismann expanded discussion about the role of cells in development and about their status as the smallest basic functional, structural and even developmental units of life, but others disagreed.

Reactions became particularly lively among American biologists around the turn of this century, especially among researchers who gathered at the Marine Biological Laboratory (MBL) in Woods Hole, Massachusetts, each summer. The first MBL director, Charles Otis Whitman, directly questioned the dominance of the 'cell standpoint' in his paper 'The Inadequacy of the Cell Theory of Development'. In particular, he regarded the cell theory of development as inadequate. The whole individual organism directs development, he insisted, for the individual has an organisation that the study of cells alone cannot explain. Whitman endorsed Huxley's view that cells are like seashells. They are left behind by the tide and map the tide's effects, but the seashells themselves do not represent or record the tidal process of change, which is ultimately the interesting phenomenon. Organisation of an individual, then, does not result simply from cell division but instead precedes and directs cell division. Whitman rejected the full implications of the Roux-Weismann theory of independent development of individual cells.

And yet he certainly did not claim that cells have no importance. In fact, Whitman inaugurated a series of cell lineage studies during the 1890s by his students and colleagues at the MBL and at the University of Chicago. These studies were designed precisely to illustrate the patterns of cellular development and the way in which the original egg takes on differentiation. Cell lineage studies consisted of using exacting histological and microscopic techniques to trace what happens to the cell and its nuclear and cytoplasmic parts as the cell divides. What happens inside, and what is the result as the cell products each begin to assume their own individualities as well? Researchers including Whitman, Wilson, Edwin Grant Conklin and a host of others carried out these detailed studies on a variety of organisms in order to assess similarities and differences which might reveal ancestral evolutionary relationships. They also sought to evaluate the extent to which the original egg cell is already differentiated by determining how regular or 'determinate' the cleavage process and products are from one individual to another. If every organism of the same type divides in the same way, and if each cell product gives rise to the same differentiated adult parts as that same cell product in other organisms, then individual development must somehow be strongly determined by the structure or content of the egg cell. If, however, there is considerable variation and flexibility, then the egg is not strictly determined, and environmental factors must direct development. Cell lineage study, in fact, focuses on cells and cell fates and reveals a complex interaction of external and internal directive factors operating on development.

Whitman insisted that it was the whole functioning organism which ultimately directs differentiation. Cell boundaries simply do not follow predetermined and invisible lines in the egg. Nor does cell division respond blindly to internal nuclear directions. Instead, cells respond, following the laws of a sort of organic physics and chemistry, to the needs of the whole, which in turn result from the long-term action of evolutionary selection. Conklin agreed, priding himself on his being 'a friend of the egg' as a whole. While some emphasised the role of either nucleus or cytoplasm in development, Conklin and others joined Whitman in stressing the significant role of both and of the integrated organism. For example, Charles Manning Child stressed that 'It is the organism – the individual, which is the unit and not the cell'.

Yet other American researchers, including E. B. Wilson who also worked closely with Whitman at the MBL, did regard the cell as basic. One could study the cell and its role in directing development and inheritance to productive effect because each cell does have significant individuality, even if co-ordinated with other cells within a more complex organism. As Wilson interpreted the revised cell theory of 1896, it held that higher life forms, whether animal or plant, consist of structural units known as cells. Out of these cells arise tissues and the other body parts. Cells are not hollow, as Hooke's choice of words would suggest, but are filled with protoplasm. The cell and its protoplasm serve as the 'physical basis of life'. Cells are all the basic, elementary units of both organic structure and function, even though they appear very diverse. In higher organisms, according to Wilson, as cells become more specialised, they enter a 'physiological division of labor'. Study of the cells and their actions must provide a basis for all study of life, Wilson felt, for:

Each bodily function, and even the life of the organism as a whole, may thus in one sense be regarded as a resultant arising through the integration of a vast number of cell-activities; and it cannot be adequately investigated without the study of the individual cell-activities that lie at its root.

The result of the co-existence of these alternative interpretations and emphases on the cell's importance even at the same summer laboratory resulted in heated discussion. This discussion, as it found expression in published lectures and papers, stimulated further attention to the cell.

The cell was generally accepted as a legitimate subject of study in its own right. And yet at least for this group of researchers, understanding of cell theory had changed. The cell could still be the most common, fundamental unit of life, but researchers did not have to take the cell as the proper unit for *all* analysis. Cells are not completely independent and do not either develop or live alone – at least most cells from multi cellular organisms do not. Cell theory and its role in interpreting development had gained considerable complexity by 1900, but this complexity complemented and refined rather than undercut the value of older cell theory. It is the historian's task to sort out the different traditions, and to assess the influence of new techniques and questions in modifying early observations, in order to explain the move from early observations of cellular structures to the cell theory and on into the developmental examinations of the late nineteenth century. So far, only parts of the story have been told, and too often from a retrospective perspective, unquestioningly seeking past equivalents of modern cells and missing such important questions as how cell theory influenced theories of development.

Cell Theory in the Twentieth Century

In this century, researchers have recognised that the most serious weakness of the cell theory is its inability in itself to explain cell-to-cell interaction or that organisation of many cells which became Whitman's stumbling-block. By mid-century, Hans Spemann and others did, however, begin to attack such interactions in various ways. Considerable recent work, especially since the Second World War, has been directed at dealing with this weakness. Studies of exchanges at cell junctions and across cell membranes have begun to show the ways in which cells join together into functional multicellular units. Advances in biochemistry and molecular biology, as well as the advent of observation techniques and equipment such as electron microscopes which allow ever more detailed resolution, have led the way in this work.

Since the Swiss physiological chemist Johann Friedrich Miescher theorised in 1869 that nuclei of cells may all consist of a characteristic substance (which he went on to identify as 'nuclein', later called nucleoprotein), biochemists have provided increasingly detailed information about the cell. While the researchers discussed above concentrated on the structural elements: on the nucleus or the cytoplasm or the whole cell and its relation to other cells, a few other researchers analysed what substance is *in* the cell. In the twentieth century, it has gradually become clear that cells of different types exhibit remarkable consistency in their cell substances. DNA material is remarkably uniform throughout all living cells, and so is much of the rest of the cellular material. This discovery has reinforced the view of the cell as the basis of all life. It has also returned biologists to the sorts of questions about evolution and the relationships among cells that earlier cell lineage researchers were asking.

If all cells contain much the same substance, as well as exhibiting similar structural elements, then perhaps the cell is the ancestral unit of life. Perhaps a simple cell actually came first, rather than a complex multi cellular and even multilayered ancestral organism such as Ernst Haeckel and most other nineteenth-century naturalists had envisioned. Such a hypothesis certainly satisfies the often-repeated call for simplicity in science.

Unicellular organisms presumably preceded multicellular organisms, perhaps (as it seems recently) with the boundary having been crossed several times. Studying cells and their similarities and subtle differences among different types of cells or between eukaryotes and prokaryotes, can provide a key to evolutionary or phylogenetic development, as Margulis 's recent investigation into *The Origin of Eukaryotic Cells* has shown. The twentieth century has thus added a new layer of significance to the original cell theory. Evolutionary study has not replaced embryology or the study of cell division or the biochemistry of cells certainly, but has complemented it. The cell theory as the statement that all life consists of basic cellular units, which arise only from other cells has gained new dimensions and new confirmation.

13

Models of Scientific Theories

Introduction

We examined a few of the popular misconceptions about scientific theories. Our remarks there were mostly negative: They indicated that scientific theories do not have certain features that many people think they have. Theories are not somehow especially "iffy" (as opposed to alleged "proven facts"), and to call a body of assertions a theory is not to suggest that the assertions are just as likely to be false as to be true. But it would be helpful if we could find some positive characterization of theories. What are they? Are they mental objects? Abstract objects? Are they not objects at all, but rather some sort of linguistic representation of external reality?

The positivists held that a scientific theory is a linguistic representation, not of external reality (which they were not sure was a notion that even made sense), but of actual and possible human experiences. Their idea was that a scientific theory is a shorthand encapsulation, in linguistic form, of the sensory experiences a practitioner using the theory either has had or would have (depending on whether it is explanation or prediction at issue) in the attempt to interact with the entities and properties the systems, if you will that the theory is about. The positivists were radical empiricists in the spirit of the British philosopher David Hume, and they saw the anatomy of scientific theories through the filter of

Humean principles. What a theory does is to capture in linguistic form the causal regularities that hold within a domain of phenomena. For this reason, the positivist model of scientific theories made much of the concept of theoretical laws, such as the Newtonian laws of motion, the laws of chemistry, and the laws of genetics. They took laws of nature, in this sense, to be representable by certain kinds of linguistic statements. Hence, it seemed appropriate to the positivists simply to treat a scientific theory as a linguistic entity, a set of sentences possessing certain special logical features. This is an approach to understanding science that seeks to capture it at its most mature and successful: when it has evolved to the point of having a mature theoretical basis. Theories, with this approach, are after-the-fact reconstructions which present an abstract picture of the phenomena relevant to the domain. The positivists were interested in the final end-product of scientific inquiry more than they were interested in the sometimes sloppy process of trial and error used to reach that final end-product. We shall see that the positivist model of that end-product was conceptually rich but fraught with technical problems from the very beginning.

Theories as Formal Languages

Logical Positivism was a twentieth-century philosophical movement primarily concerned with theory of knowledge (epistemology), mathematical logic, and philosophy of science. Positivism was born in central Europe, but the flight of European intellectuals away from the growing menace of Nazi fascism in the years immediately preceding World War II spread positivism's influence across the larger part of the English-speaking world. Many of positivism's founders were initially trained as physical scientists before they became philosophers. For example, Rudolf Carnap was trained as a physicist, but he eventually came to occupy a position of enormous influence and respect within the positivist movement.

The logical positivists had a peculiar "spirit," an overriding set of intellectual and philosophical attitudes, preferences, and commitments. They got the second part of their name from the work of the nineteenth-century French philosopher August Comte. Comte had speculated that human intellectual culture evolves through three stages: theological, metaphysical, and positive. In the theological stage human science is at its most primitive, awash in religious superstition and busy attributing magical person like properties to what are actually inanimate processes in nature. The world is populated by deities of various kinds who are held to cause what happens. In the metaphysical stage the magical gods are replaced by impersonal forces such as gravity and charge. This is still an error, according to Comte, but a better sort of error because it deanthropomorphizes the universe. Nevertheless, these forces are conceived as mysterious and occult in nature, "hidden" away "inside" things. The third and final stage is the positive stage in which the metaphysical forces are done away with in favor of commonsense observational concepts. The positivists followed Comte in being generally antireligious. They also advocated the demystification of nature. They adopted a radically empiricist orientation toward most philosophical issues (don't try to "think" the answer via metaphysical insight, let sensory experience decide the question, whatever it is). Many of them doubted the existence of disputed abstract entities (such entities don't really exist), or else they rejected outright all metaphysics, taking over from Comte the practice of treating 'metaphysics' as a dirty word in philosophy. They practiced a healthy skepticism toward the inflated claims made by theorists in many special fields of inquiry. Most important of all for our purposes, the logical positivists had a specific methodology that they employed in the analysis of philosophical problems, a methodology that explains the first half of their name, one in which mathematical logic played the lead role.

Mathematical logic was developed originally by Gottlob Frege and Bertrand Russell over a period spanning roughly from 1870 to 1915. Why were the positivists so smitten with mathematical logic? The answer is, because they understood mathematical logic to be a universally applicable, maximally precise cognitive tool, a tool that could be used to analyze successfully a host of abstract, difficult-to-understand concepts, problems, and issues. And a successful analysis ends our mystification about whatever it was that puzzled us by showing us how to understand it in terms of other things that we already understood previously.

The universal applicability of mathematical logic is due to the fact that it is a *subject-neutral language* that can be used to capture and express the claims of any special subject matter. And the precision of mathematical logic is due to the fact that the main symbols of this subject-neutral language can be provided with *mathematically precise and unambiguous meanings*. In short, you can use mathematical logic to talk about anything under the sun, and the claims you produce with it each have a unique and computable meaning. This is a splendid outcome because, as generality and precision increase, confusion and ambiguity decrease. It is no surprise then that this relatively new "toy" of mathematical logic was taken by the positivists to be a kind of philosophical wonder drug with almost Herculean powers.

The commitment to the power and predominance of mathematical logic was especially influential in positivist philosophy of science. The core assumption in positivist philosophy of science was this: Mathematical logic contains enough power and precision to allow us to capture and express in a useful and understandable way the most important causal and explanatory relationships that occur out in the real world, that is, in the domain of investigation in which a given scientist works. An example will help here. For almost 400 years since the Early Renaissance, philosophers of science (and philosophers in general) have been fond of asserting that the events that take place in the natural world are organized and controlled by laws of nature. The laws of nature that are operative within a given area of the natural world are perhaps the single most important feature of that area to the scientific discipline that investigates it. This was certainly true for the positivist model of science; for, under the positivist model of science, laws of nature had two important roles to play. First, laws of nature are what fix the causal relations in a system under scientific study: They are responsible for the relations of causation which hold between the subparts of a system and between that system and any other systems. Second, laws of nature make possible the explanation of natural phenomena. A change of properties in a system is explained by showing how that change was necessitated by the relevant laws of nature acting on the initial conditions in the system. What is of interest to us here is the fact that laws of nature play such an important role in science, and yet saying just what a law of nature is proves to be rather difficult. Is a law of nature a command of God's? Or is it a mysterious connection of "necessity" between material properties? Prior to the rise of the positivist model of science no philosopher of science had succeeded in offering a clear analysis of the concept of a law of nature. The only clear conception that the metaphysicians could supply was theological, but a theological account of laws of nature was totally inconsistent with the spirit of positivism noted above.

The positivists had a solution to the puzzle, a solution that assumed that mathematical logic could rescue us from the vagueness surrounding the notion of a law of nature. The positivist solution was to treat laws of nature as representable by material conditionals of universal scope. A material conditional is a symbolic sentence in mathematical logic that has a specific logical form. In logic textbooks it is standard practice to claim that the material conditional represents what in English we say by using an 'If . . ., then . . .' pattern of speech. 'If A, then B' is a material conditional in which the truth of the component statement A is a condition sufficient for the truth of the component statement B. Recall that a vital stage in the generation of an allergic reaction is the physical binding of the allergen molecule to the antigen-binding portion of E-class antibody molecules, otherwise known as immunoglobulin E (IgE). Immunoglobulin E is one of five general classes of immunoglobulin antibodies found in the mammalian immune system. Immunoglobulin-E molecules have the same general three-dimensional structure as molecules of immunoglobulin classes A, D, and G, except for one notable difference: IgE molecules each have five distinct amino-acid regions in their heavy chains, whereas molecules of the other three classes of immunoglobulins mentioned each have four distinct amino-acid regions in their heavy chains. To oversimplify, IgE antibodies are structurally longer than either IgA, IgD, or IgG molecules. This is a law of immunology. It is structurally essential to IgE molecules that they each have five-component rather than four component heavy chains. The positivist model of science would analyze this law of immunology as a universally quantified material conditional; that is, it would translate the English version of the law.

All IgE molecules have five-component heavy chains into the symbolic sentence of first-order mathematical logic. 'Eig' is the symbolic term that refers to the property of being an IgE molecule, and 'H5' is the symbolic term that refers to the property of having five-component heavy chains. The idea here is that the symbolic statement is a precise logical equivalent of the English one above it. The symbolic statement would be read in English as 'For any object in the entire universe call that object x, if x is an IgE molecule, then x has five-component heavy chains'.

To take another example from chemistry, the law that all pure metals conduct electricity can be rendered symbolically by the universally quantified conditional.

Now one might Justifiably ask why this is something to get excited about. The positivist can answer that because we understand with mathematical precision the meaning of the material conditional and the meaning of the universal quantifier , we now know with much more precision than we would have if we stayed with ordinary natural English just what the relationship is between being an IgE molecule and being an object with five-component heavy chains: We have arrived at the central assumption behind the positivist model of science. *The positivists assumed that the meaning of logically defined symbols could mirror or capture the lawful relationship between an effect and its cause.* Or, to put it in more general terms, the positivists assumed that logical relations between formal symbols could accurately represent external relations between the objects, properties, or events to which the formal symbols referred.

Now that we have gotten our feet wet in the sea of logical positivism, we are ready to dive in and take a look at the full positivist model of science. As we discussed above, the positivist model took theories as its unit of analysis in philosophy of science. Science is a collection of theories, according to the positivist, so we will want to see the positivist model of a scientific theory. A scientific theory is a formal language, a language expressed in first-order mathematical logic, which meets certain special conditions. First, the purely logical relations that are expressible in the language are the five standard truth-functions,

symbol	rough English meaning
v	or
&	And
?	if...,then ...
~	Not
=	... if and only if...

plus the two standard quantifiers,

? for all objects ...

? there is at least one object ...

and the identity sign needed to assert that two referring expressions refer to the same thing,

= ... is identical to ...

Second, the theory may contain any number of purely mathematical symbols such as 'Ó' that may refer to various constants (Planck's constant, Avogadro's number, and so forth). Third, all remaining symbols in the theory are neither purely logical nor purely mathematical symbols. They are what we might call substantive symbols, symbols like 'E^{ig}' which refers to the property of being an IgE antibody molecule. Fourth, these substantive symbols are divided into two completely separate kinds. Some of the substantive symbols refer to observable objects, properties, or events, and they are called observational terms. The remaining substantive symbols refer to non observable more commonly called theoretical objects, properties, or events, and they are known as theoretical terms. Objections have been made to this manner of speaking, because the proper contrast term to 'observational' is presumably 'non observational' rather than 'theoretical'; but we will stick with the standardly used terms throughout this text, thereby avoiding both a dispute over a technicality and terminological confusion. Fifth, the core content of a scientific theory is contained in the set of its purely theoretical laws of nature. These will be universally quantified conditionals whose only substantive symbols are theoretical terms. Sixth, but most important for our present purposes, the positivist model required that *every theoretical term in a scientific theory must be provided with an explicit definition composed entirely of observational terms*. Such explicit definitions were called correspondence-rules, or c-rules. For example, a positivist allergist has no business talking about the $F(ab)_2$ portion of an antibody molecule, which is the Y-end of it capable of binding antigen (the general term for any molecule bindable by an antibody), unless it is possible to provide an explicit definition of the theoretical term '$F(ab)_2$' in which the defining symbols are all observational terms. Now, to require that the definition be an *explicit* one is to require that the observational side of the definition be such that whenever it truly applies to an object, property, or event, its application is necessary and sufficient for the theoretical side of the definition to be truly applied to that object, property, or event. Put another way, for two terms to be necessary and sufficient conditions for each other means that whenever the one term applies to a phenomenon so does the other, and whenever one of the terms does not apply to a phenomenon neither does the other. The logical relation that holds between two terms when each is necessary and sufficient for the other is represented by the logical connective called the material biconditional, '='. So, our positivist allergist must provide a c-rule for the theoretical term '$F(ab)_2$' that has a particular logical form: It is a symbolic sentence in first-order logic that is a universally quantified material biconditional.

Suppose our positivist allergist decides to define observationally '$F(ab)_2$' in terms of a laboratory procedure combining a pepsin digest with gel immuno electrophoresis. This standard laboratory procedure can be performed on blood serum to observe the various fragments or portions of antibody molecules, including the $F(ab)_2$ fragment. If the immunologist makes no mistakes in the procedure, the $F(ab)_2$ portions of the antibodies show up as visible arclike bands against a uniform background at the end of the procedure. The figure shows an immuno electrophoresis gel in which an antigen well and an antibody trough have been cut. The arc to the right of the well between the well and the trough was caused by the reaction, of the $F(ab)_2$ fragments with their corresponding antigen. Those deeply theoretical objects have made contact with human observational space at that endpoint in the procedure. So, says the positivist allergist, let us simply define '$F(ab)_2$' in terms of this procedure that results in the observation of the visible arcs. Let us represent the observational procedure used to generate the bands by the formal 'O^1'.

In ordinary English the above symbolic sentence translates as, 'anything is an $F(ab)^2$ portion of an antibody if and only if it produces observable arcs of a certain kind at the end of observational procedure O^1'.

The ability to provide c-rules in the form of explicit definitions served a doctrinaire purpose for the positivists. Many positivists claimed that the c-rule requirement was what kept science "honest" in the sense that it served to inhibit scientists from arbitrarily making up exotic theoretical objects, properties, or vents.

The c-rule requirement therefore was one main feature of a scientific theory that distinguished genuine scientific theories from their phony imitators. A phony science is called a pseudoscience, and many such pseudosciences are big businesses in contemporary life. Examples are not hard to find. Astrology and psychic healing, for example, are big moneymakers for their practitioners. One interesting aspect of pseudoscientific theories is that they contain no c-rules for their pet theoretical terms. It is unheard of for an astrologer to provide a c-rule for 'astral influence' or for 'influence of the ascending house'. The c-rule previously provided for '$F(ab)_2$' is admittedly very esoteric and unfamiliar to those unacquainted with immunological science, but at least it can be provided. That is more than can be said for the typical pseudoscientific theoretical term.

The reader might now ask why it is such a negative feature of a theory that it cannot provide c-rules for its theoretical terms. Why couldn't the pseudoscientists defend their theories with a simple shrug of their shoulders and a cavalier "so what"? The reason why such a defense would be lame, at least according to the positivist, is because *c-rules tell us how the alleged theoretical entities defined by them make a difference to observational experience.* If you cannot find a difference that the alleged existence of the theoretical object, property, or event makes to the observable world, then the positivist would suggest the absence of any good grounds for supposing that such theoretical objects, properties, or events exist at all. The positivist model tied the grounds for asserting the existence of a thing to the difference its existence would make to our observational experience. Suppose a controversial immunologist proclaims the discovery of a new class of theoretical objects that are active in the mammalian immune system at an extremely fine level of detail. This immunologist invents a new theoretical term to name these objects: 'ferple'. Are there really ferples? we wonder. So we question this immunologist about ferples only to find that ferples are so darn theoretical, so darn small, and so on, that they can never be detected by any observational means. They exist, insists the maverick immunologist, but we can never observationally experience their presence or any of their direct causal effects. The only causal effects ferples have that we can observe are so far downstream from ferples themselves involve so many intermediate causal links that it is just as likely that ferples are not really the first link of that chain of causes and effects as it is likely that they are the first link. The problem is not that there is currently no possible evidence *for* the existence of ferples; as we have imagined the case, the problem is that there is no possible evidence that would count *against* the existence of ferples. Now if we are positivist philosophers of science, we must reject the existence of ferples. They make no difference to the way the world can be observed to be by us, so we have no warrant for supposing they exist. To put it bluntly, if they make no difference to observational experience, then they aren't there! Another way to understand the point the positivist is making is to consider what our estimate would be of the scientific worth and meaning of a theoretical term whose use is consistent with any and all possible observational experiences. The positivist would argue that such a term is scientifically meaningless. It is important for the reader not to misunderstand the positivist here. The positivist is not arguing that science shouldn't be allowed to postulate theoretical entities; the positivist is saying only that any such theoretical entities must be given an explicit market value within observational experience.

Requiring that theoretical entities have an observational market value was supposed to help reduce the number of bogus theoretical entities in science. Whether the requirement was ever capable of achieving that outcome is doubtful, but we can at least understand how it was supposed to work in broad outline. The history of science contains examples of theoretical entities and properties that turned out in the course of inquiry not to exist at all. For example, early in the history of chemistry the combustion of a chemical substance was postulated to be due to the release of something called phlogiston, a mysterious "principle" of pure flammability often conceived to be a substance of some sort that all combustible bodies contained. There is no such substance or principle. In physics, once upon a time, space was held to be everywhere pervaded by a mysterious nonmaterial substance called the aether. Aether was the alleged "medium" through which energy propagated and ordinary matter moved. The case of aether is justly famous in the philosophy and history of science for being at the center of the overthrow of Newtonian physics at the beginning of the twentieth century. Again, there is no such aetherat least there is not if Einsteinian physics is correct. The positivists were aware of such historical cases of bogus theoretical entities and hoped that the c-rule requirement in their model of science would contribute somehow to lessening the frequency of bogus theoretical terms; for, if necessary and sufficient conditions expressible solely in the observational vocabulary are required for every theoretical term, the stringency of such a requirement would surely serve as a deterrent to the casual postulation of theoretical entities.

Dummy Scientific Theory to Illustrate the Positivist Model

The discussion of the positivist model of scientific theories up to now has been a bit abstract. A specific example would help us to understand how the model was supposed to work. Let us construct a crude and completely phony positivist theory of allergic disease so that we might illustrate how the positivist model was meant to work. I will call this theory a "dummy" theory because it is completely made up and because it bears the same sort of relation to genuine immunological theories of allergic disease that human-shaped dummies (like the dummies used in car-collision tests) bear to living humans. What we lose through the theory's phoniness we gain back from its simplicity and clarity. We will be able to see all the components of the positivist model in sharp relief.

Our dummy theory of allergic disease contains four substantive theoretical terms, three theoretical laws of nature, and four c-rules. We will allow the theory to contain any number of additional observational terms, and we shall assume for present purposes that telling the difference between theoretical terms and observational terms is unproblematic. Since the positivist model took the core component of a scientific theory to be its laws of nature especially its purely theoretical laws of nature we will begin with the three laws in the dummy theory of allergic disease. I will present each law in both its English and its symbolic form. The laws will be numbered L_1 through L_3.

English	Symbolic
L_1: All mammals with gene G have serum blickledorps	(?x)(Gx ? Bx)
L_2: All mammals with serum blickledorps secrete microspam	(?x)(Bx ? S^mx)
L_3: All mammals that secrete microspam experience allergic symptoms	(?x) (S^mx ? Ax)

The properties of being a mammal with gene G, having serum blickledorps, being a mammal that secretes microspam, and being a mammal who experiences the symptoms of allergy, represented symbolically by the theoretical terms 'G','B', 'Sm', and 'A', respectively, are all theoretical properties. It is important to notice that I have deliberately chosen to use a couple of phony theoretical terms like 'blickledorps' and 'microspam' (there actually are no such things) to emphasize by exaggeration their theoreticalness. This is in the spirit of positivism. As far as the positivist is concerned, you don't need to know what 'blickledorp' and 'microspam' mean on a conceptual level (that is, their sizes, colors, charges, polarities, and such, if any); you only need to know the meanings of the *logical* relations that hold between 'blickledorp', 'microspam', and any other terms, be they theoretical or observational, with which they are connected by the laws of the theory. Here we see hard at work the positivist assumption mentioned previously, the assumption that logical relations are capable of accurately representing causal connections out in the real world. For example, according to L_2 a mammal with serum blickledorps is an entity that necessarily secretes microspam. According to L_1, having serum blickledorps is a causally necessary result of the possession of a certain gene. How do we know this? The positivist answers: because the logical relations between the linguistic symbols 'G', 'B', and 'Sm' in laws L_1 and L_2 represented by the material conditional accurately capture or mirror the objective causal connections between physically four-dimensional genes, blickledorps, and microspam molecules. To the positivist, the logical relation denoted by '?' is less cognitively remote and better understood (because it is precisely and unambiguously defined in mathematical logic) than the objective causal connections themselves; so, we can be more sure of the logical relations than we can of the objective causal connections. It is a question the reader should ask at this point, whether this general assumption by the positivist that *logical connectives mirror causation* is warranted or not. It remains for us to complete our dummy theory of allergic disease by providing c-rules for our four theoretical terms. To facilitate understanding, I will leave the theoretical side of each c-rule in its formal version but present the observational side of each c-rule in English.

1. (?x) (Gx = x's fingernails produce green spots when pressed continuously for five minutes)
2. (?x) (Bx =pricking x's skin immediately produces a bright turquoise spot on the surface of x's tongue that lasts four seconds before fading away)
3. (?x)(S^mx = x's breath smells like maple syrup)
4. (?x) (Ax = x presents with the typical symptoms of allergieswheezing respiration, watery eyes, swollen and inflamed nasal tissues with pink coloration, hives, and so forth)

We should be clear that the right side of each of these c-rules (the side in English) is to be taken as a purely observational description. Now that we have our illustrative dummy theory completed we may ask to be shown what benefits flow from logically formalizing a theory in such a manner. The positivist answers that there are important benefits forthcoming. Perhaps the most prominent benefit is that *theoretical inferences become a matter of making deductively valid inferences*. This makes such inferences completely clear, precise, and transparent to the practitioners who use the theory. Suppose, for example, that a certain patient attended by our positivist allergist is examined by the allergist and tested for blood blickledorps. This patient's blood is found to contain no blickledorps (assume the test for serum blickledorps is extremely accurate). Now, by law L^1 and the deductive inference rule logicians call *modus tollens* (which stipulates that from a conditional statement 'if A, then B' and the statement 'not B', we can validly infer the statement 'not A'), the allergist can infer deductively that the patient does not have gene G. So what?, you say, why is that a big deal? The positivist answers that it is a big deal because this inference is not a guess, not a vague hunch or prediction; rather, it is a statement the allergist can proclaim with *deductive certainty*. To say that a pattern of inference is deductively certain is to say that *if* the premises (the supporting statements, in our example they are the two premises L^1 and the statement 'the patient's serum contains no blickledorps') are true, *then* the conclusion (the inferred statement, in our example it is the statement 'the patient does not have gene G') *must be* true also. This alleged elimination of guessing from theoretical inference is one major advantage the positivist claims for the positivist model of scientific theories. Further, the *modus tollens* inference counts as a double-check against any independent test for having gene G; that is, we can infer the lack of gene G with out having to administer any independent tests for G to the patient.

A second major advantage of logically formalizing a theory should perhaps be mentioned at this point. Under the positivist model, an especially powerful and precise model of scientific explanation becomes available to us. Using our dummy theory, for example, the appearance of the symptoms of allergy in a human patient could be scientifically explained in the sense that a statement describing the patient's symptoms can be logically deduced from L_1, L_2, and L_3 together with an additional pertinent premise that attributes to the patient in question the gene G mentioned in L_1. Explanation in science, claimed the positivists, consists in showing how the event to be explained was necessitated by the relevant laws of nature together with the relevant initial conditions obtaining in the system in question. Thus, explanation in science is put on a solidly precise footing. Scientific explanation need not remain the largely impressionistic and vague notion it was and often remains outside the positivist model.

The positivist model of scientific theories dominated philosophy of science for nearly forty years after its introduction in the 1930s. The precision and power of the positivist model silenced both its critics outside the positivist movement as well as any self-doubters within the positivist camp.

14

Reductionism, Antireductionism and Supervenience

Introduction

Miletus was a commercial seaport in a region known to the locals of the time as Ionia. Around 600 B.C.E. a great sage lived in Miletus whose name was Thales. Among other achievements attributed to this wise person was the prediction of an eclipse of the sun that took place in the year 585 B.C.E. (scholars believe that Thales predicted it to the year, not to the daya still impressive degree of precision given the primitive state of organized astronomy at the time). Thales was the first scientist. He is traditionally considered the first philosopher as well. Why was Thales the first scientist? Because he tried to provide an account of the universe in entirely "natural" terms. He is said by Aristotle to have claimed that the *arche* (first principle) of all things is water, meaning that all things are made of water, that all things are water. This was something brand new at the time, for what was then the traditional manner of explaining most things involved making appeals to supernatural deities and magical forces of various kinds. Far from being a silly hypothesis, Thales' theory brought theoretical speculation about the universe down from the supernatural realm and into the world of ordinary human experience. Some philosophers argue that Thales had rational grounds for choosing water as the basic "stuff" of the world. Water was in ancient times the only natural substance known to exist in all three states of matter gas, liquid, and solid. Further, all forms of life that the ancients knew about required water to live and reproduce. We do not know if Thales chose water as the basic stuff of the world for such reasons as these, but it certainly would be in character for him to have used such reasoning given his preference for calling on natural causal forces to explain things. Further, Thales' theory was maximally *unifying* in the sense that it sought to understand qualitatively different phenomena as being due to the same one underlying reality: water. That sort of unification of separate phenomena under a small number of basic explanatory concepts is one feature that distinguishes science from most forms of speculative tribal wisdom.

What Thales began other Milesians continued Anaximander (a-naks-e-MANder) was another Milesian scientist and philosopher, only he claimed that all things stem from t*o apeiron,* an untranslatable Greek phrase that usually gets rendered as 'the boundless'. The boundless might sound mystical and otherworldly to trendy modern cars, but Anaximander meant nothing especially supernatural by the term. It was a kind of neutral stuff, without magical powers, and it was certainly not a deity of any kind. Anaximenes (an-acks-SIM-e-knees), who came after Anaximander, holds an honored place in the history of science because of a vital improvement on Thales' and Anaximander 's theories for which he was responsible. Thales had not provided a causal mechanism by which water gets transformed into things that appear not to be water. A piece of granite rock surely does not appear to be very waterlike. Hence, how can it be true that it is made of water? Thales apparently had nothing to say about this issue (or, if he did have something to say, we have lost the record of it). Anaximander had more to say on how the boundless is transformed into ordinary material objects, but his explanation did not involve a causal mechanism. The boundless is broken into parts in an act of "injustice," for which justice must be meted out by the eventual contraction of everything back into the boundless. Here one still sees a holdover from anthropomorphic modes of understanding the universe. Anaximenes, on the other hand, provided the first *naturalistic* causal mechanism by which a substance that possesses one bunch of qualities can be transformed into a substance with a different bunch of qualities. Anaximenes said that all things stem from air (not water or the boundless), and that all qualitative *differences* among things are due to different degrees of rarefaction and condensation of the air out of which all things are made; so, all qualitative differences are reduced to quantitative differences. Here was science to be sure. For the main job we expect science to do is to tell us, in non magical terms, what is really going on in the world what there is in it, how what there is in the world got that way, and where what there is in the world will go next.

Twenty-five hundred years later we still expect science to tell us what is really going on in the universe. Some philosophers of science claim that how we measure progress in science comes down to the degree to which the several sciences are unified into one interconnected network of theories, a network that indicates that the universe is a cosmos and not a chaos, that it is in a nontrivial sense a single thing whose parts fit together into a coherent pattern. The term for the relationship between scientific theories that the positivists thought made possible such an interconnected network of theories is **reduction.** To reduce one theory to another theory was, under the positivist model of science, the ultimate goal of science. A reduction is naturally interpreted as achieving progress in an ontological sense. If, for example, chemistry is reducible to physics, as indeed it seems to be, then if we were to make an ontological inventory of all the kinds of objects and properties there are in the world, we would not in addition to all the physical kinds of objects and properties (that is, all the kinds of physical subatomic particles and their subatomic properties) need to list any chemical objects and properties. Why not? Because chemical objects and properties just are combinations of physical objects and properties. For example, the chemical property of *having a valence of* + *1* is "really" just the physical property of being a single-electron donor, that is, of how filled the outermost electron shell of the atom is. Chemical valence is not something distinctly chemical and nonphysical that must be added to our inventory. Instead of there being more basic objects and properties in the universe, there are fewer. And that is good, because the fewer the basic objects and properties, the lower our risk of error in doing science (there is less for us to be wrong about). The last sentence leads naturally to a principle of scientific method often called Ockham 's Razor after its most famous advocate the medieval philosopher William of Ockham. Ockham 's Razor holds that we should never multiply the number of independent entities and properties in our scientific theories beyond the minimum necessary to account for the phenomena in question, that we should never prefer a theoretical account of the universe that represents the universe as bloated with unnecessary entities and properties to a theoretical account of the universe in which it is a thin universe with just enough in it to appear as it appears to us. Ockham 's principle can be used like a razor to pare or trim away from our scientific theories unnecessary posits and concepts, hence the principle's rather unusual name.

There have been a few successful reductions in the history of science. Thermodynamics in physics is reducible to statistical mechanics. In biology Mendelian genetics is reducible, or so some have claimed, to molecular genetics. The reduction of chemistry to physics has already been mentioned. There have also been a few what we might call partial reductions, reductions that have not come close to being completed but which have met with enough partial success to encourage further pursuit of a reductive research strategy. Some aspects of psychiatry seem reducible to neurophysiology, for example, so perhaps all of psychiatry is reducible to neurophysiology if we pursue that research goal with enough patience and stamina. Some proposed reductions have met with considerable controversy. The sociobiologists aim to reduce sociology to evolutionary biology. Needless to say, most sociologists are not especially fond of this proposal. In the 1970s some theorists sought to reduce cognitive psychology to neurophysiology plus the theory of artificial intelligence. This raised many hackles throughout the community of psychologists. It is interesting to note that this was one area of intense debate that had a direct impact on the philosophy of science. We will explore that impact in section 5.4. For now we need to ask the obvious question: precisely how is a reduction carried out? What is involved in reducing one theory to another? Different models of intertheoretic reduction have been offered to answer this question.

Nagel's Model of Reduction

The philosopher of science Ernest Nagel occupied a prominent position among American philosophers for nearly thirty years. He trained several outstanding philosophers of science, of more than one generation, during his long tenure at Columbia University. Nagel was trained in the positivist model of science. He never completely abandoned it, but he nevertheless moved away from a strictly positivist approach in many respects. His monumental text The Structure of Science, published in 1961 (one year before Thomas Kuhn's anti positivist masterpiece appeared), was the culmination of the neopositivist attempt to salvage some essential core of positivist doctrine at the center of philosophy of science, a particularly heroic attempt in the face of the then newly arising historicist model of science that was eventually to be associated with Kuhn and his aftermath. Nagel's book is a bit imposing: thorough, technical, dry at times, but clearly written and written by someone who knows the philosophy of physics inside and out, the text presents many discussions that have since become classics of their respective subject matters. One of those classic discussions is the famous chapter on reduction, in which Nagel presented a model of reduction that was applauded for its clarity and precision even as it was skewered for its quixotic rigidity and severe formalism.

Nagel's model of reduction was tied to the positivist conception of scientific theories in the sense that reduction was applicable only to theories that were formalizable as first-order languages in the way the positivist conception demanded. A theory, in order to be reduced or to do the reducing, must, for Nagel, be expressible as a set of formalized sentences with the sort of logical structure. The theory must have a set of core theoretical laws, and it must have a vocabulary distinguishable into two mutually exclusive classes of terms: theoretical and observational. Assuming that scientific theories can be so formalized, Nagel's model of reduction is remarkably clean in principle but extremely sloppy to pull off in practice.

Let T_1 and T_2 be formalized theories. Then T_2 reduces to T_1 if and only if the following conditions hold:

1. The Condition of Connectability: For every theoretical term 'M' that occurs in T_2 but not in T_1, there is a theoretical term 'N' that is constructible in T_1, but not in T_2 such that for all objects x, x has M, if (and possibly only if x has N.

The sentence isolated immediately above is called a bridge law, because it serves as a bridge to link two distinct linguistic expressions in two distinct theories. The requirement represented by the condition of connectability is that a bridge law must be provided for every theoretical term in the reduced theory that is not in the reducing theory. It is important not to misunderstand the condition of connectability. The condition does not require that 'N' mean the same as 'M' means. Nagel is careful to say that 'N' can either be coextensive with 'M' or else a sufficient condition for 'M'. Of these two possible relations, coextensitivity is the stronger, but it is not equivalent to sameness of meaning. To see the important difference between sameness of meaning and coextensivity, an example will be helpful. Researchers believe that the gene associated with the disease cystic fibrosis resides on chromosome 7. Suppose this is in fact the case, and suppose further that eventually the branch of specialized medicine that treats diseases like cystic fibrosis is reduced to some form of supergenetics. That would mean, if connectability was taken to involve coextensitivity, that there will be some bridge law to the effect that, for example, for all organisms o, o has the cystic fibrosis gene if and only if o has DNA nucleotide sequence Z-2134 at location X on chromosome 7.

But clearly the expression 'cystic fibrosis gene' does not mean the same as the expression 'DNA nucleotide sequence Z-2134 at location X on chromosome 7'; for the latter expression mentions a chromosome and specifies a long sequence of nucleotide triplets and the former expression does not. The two expressions have different meanings, but they nevertheless are expressions predicable of all and only the same things, that is, of exactly the same things. Any organism to which one expression is applicable is an organism to which the other expression is applicable, and whatever organism one is not applicable to, neither is the other applicable to that organism. The technical term for expressions that are applicable to exactly the same things is 'coextensive'. To take another example, the two terms 'renate' and 'chordate' are coextensive. A renate is a creature that possesses a kidney, and a chordate is a creature that possesses a heart. It is simply a brute regularity of nature that every creature which has one has the other there is no creature that possesses one but lacks the other. But, obviously, the two terms don't mean the same at all, for the meaning of each term involves a different type of bodily organ.

What Nagel's condition of connectability demands is that, for every theoretical term in the theory to be reduced that does not occur in the reducing theory, we would ideally want there to be some coextensive expression constructible using the vocabulary of the reducing theory, although short of that we will settle for an expression constructible in the vocabulary of the reducing theory that denotes a sufficient condition for the application of the reduced term. A good example would be the chemistry case mentioned above involving valence and electron donation. The predicate expression 'being a single-electron donor' is coextensive with the expression 'having a valence of + 1' (not synonymous with it). Now once we have all the bridge laws constructed we must meet one further condition to reduce T_2 to T_1:

2. The Condition of Derivability: It must be possible to derive logically all the theoretical laws of T_2 from the theoretical laws of T_1 together with the bridge laws. Let 'B' be the conjunction of all the bridge laws constructed in step 1, and let 'L_2' and 'L_1' be the conjunctions of the theoretical laws of T_2 and T_1, respectively. Then the Condition of Derivability requires that: (L_1 & B) ? L_2.

The reader is reminded that the symbol '?' represents the material conditional in first-order mathematical logic.

There is a refreshing clarity and precision to Nagel's model of inter theoretical reduction. The two steps have an intuitive appeal: First we must logically connect the terms of the respective theories by constructing bridge laws; then we must be able to derive logically the laws of the reduced theory from the laws of the reducing theory plus the bridge laws. What could be more elegant and to the point than that? It was a model that Nagel showed made sense of certain previously obtained reductions in physics, for example, the reduction of thermodynarmics to statistical mechanics. But there were a couple of glitches in the otherwise rosy picture of reduction that Nagel had painted. One glitch is relatively minor and the other glitch is much more serious.

In the usual case, T_2 is not reducible to T_1 as T_2 stands, for T_2 is technically false as it stands. The reduced theory is often false in the sense that it only holds good in special circumstances. The reducing theory is the correct general account of the phenomena in question. For example, suppose it were possible to derive logically the laws of Newtonian physics from the laws of Einsteinian physics plus a suitable conjunction of bridge laws.

For Newton's laws are technically false, and so anything that logically implied them would be false also (in virtue of our old friend the logical rule of inference, modus tollens). Surely we cannot blame the bridge laws, given that they were specially and carefully constructed for the reduction in question. Therefore, we must take the reducing theory to be false. This is clearly an unwanted result. The solution to the quandary is to correct the errors in Newton's laws first; that is, we must correct the errors in Newtonian physics before we reduce it to Einsteinian physics. Physicists often say that Newtonian physics is just a "special case" of Einsteinian physics; Newtonian physics is what you get when you take the Einsteinian laws and apply them to systems whose velocities are well below the speed of light and whose masses are well below that of a stellarsize object. Well, yes and no. Yes, if you first "correct" Newtonian physics (for example, you must make mass dependent on velocity, and you must remove instantaneous causation at a distance). No, if you fail to correct Newtonian physics but instead take it as it is. But then we are caught in the following dilemma. If we correct the theory to be reduced beforehand, then we don't really reduce it, but only some corrected version of it; in our example, we don't really reduce Newtonian physics to Einsteinian physics, we reduce "Snewtonian" physics to Einsteinian physics (where 'Snewtonian' refers to what is really a different theory from the original Newtonian physics). On the other hand, if we fail to correct the theory to be reduced beforehand, then the reduction shows that the reducing theory is false (for it logically implies something false). So, either we failed to reduce the original theory we wished to reduce, or else the successful reduction in fact destroyed the truth of the reducing theory. Either we fail or else success is logically deadly. Neither result is a happy one for the reductively minded practitioner. In fact, the result is sufficiently unhappy that it spurred philosophers of science to construct a separate model of reduction.

Kemeny and Oppenheim's Model of Reduction

Two philosophers of science, John Kemeny and Paul Oppenheim, argued that the Condition of Connectability in Nagel's model of reduction is almost never fulfilled in what are otherwise considered successful reductions. The reductionist is almost never able to construct term-by-term bridge laws, for the holistic interdependence of theoretical terms defeats the attempt to do so. The bridge laws become so huge and unwieldy as to be useless, if not meaningless. So, Kemeny and Oppenheim built a model of intertheoretic reduction that contained neither of Nagel's conditions; for you can't meet the Condition of Derivability without having first successfully met the Condition of Connectability (you can't logically derive the laws of the reduced theory minus bridge laws that connect the two theories in question). Kemeny and Oppenheim argued that what matters in a reduction is that the reducing theory be able to explain any observational data that the reduced theory explains, that the reducing theory be in some genuine sense different from the reduced theory, and that it be at least as "systematic" as the reduced theory. Their model also did not try to provide necessary conditions for a successful reduction, but only sufficient conditions. Again, let T_1 and T_2 be formalized theories. Then, under the Kemeny and Oppenheim model of reduction, T_2 is reducible to T_1 if

(1) T_2 contains terms not in T_1

(2) any observational data explainable by T_2 are explainable by T_1

(3) T_1 is at least as well systematized as T_2.

Condition (1) ensures that there is an ontological point to reduction. Condition (2) makes explanatory power a key factor in the evaluation of intertheoretic reductions. Condition (3) is required, said Kemeny and Oppenheim, to rule out "wrong-way" reductions. For example, reducing psychology to astrology is forbidden because astrology is theoretically more disorganized, more unsystematic, than psychology.

The fatal flaw in the model is the third condition. It would certainly appear to be the case that some reductions involve reducing a simpler that is, a better systematizedtheory to one that is more complicated and less well systematized precisely because the reducing theory doesn't oversimplify the target phenomena where the reduced theory does oversimplify the target phenomena. Consider the reduction of chemical theory to quantum physical theory. This is surely a reduction going in the correct direction; yet, it is just as surely a reduction of a fairly well systematized body of theory to one that is notoriously, if not infamously, conceptually problematic and simply not as well systematized. Are fundamental subatomic entities particles or waves? They are both, says Niels Bohr? But a wave is an entity spread across more than one point of space time and a particle is an entity not spread across more than one point of space time, we object. How is this moving from a fairly well systematized macroscopic chemistry to a theory of quantum phenomena at least as well systematized? The answer is that, of course, at present it is not any such movement at all; for the history of quantum physics up to now suggests that the ultimate microstructure of nature is inherently indeterministic and sloppy (there are currently over 200 distinct subatomic wavicles), and on those grounds alone quantum physics may be doomed to experience a kind of permanent upper limit on how well systematized it can become a limit surely exceeded by many of the higher-level theories we seek to reduce to it. Of course, this has not discouraged quantum physicists from seeking to unify quantum phenomena through reduction to some small set of fundamental objects and properties, but the point is that methodologically they have no grounds for being guaranteed up-front that their attempts at unification must succeed.

To take an example closer to our chosen domain of inquiry in this text, consider the recent attempts to begin reducing what were previously thought to be unrelated diseases to a theory that treats them all as autoimmune diseases. My prediction is that in 50 to 100 years autoimmune medicine will be a separate specialty. It will achieve such an independent status by "stealing" for its focus of inquiry phenomena from a number of currently distinct fields of medicine. The following is a list of some of the diseases known to be autoimmune in nature and the medical fields that are currently responsible for studying and treating them.

1. systematic lupus (rheumatology)
2. insulin-dependent diabetes (endocrinology)
3. myasthenia gravis (neurology)
4. multiple sclerosis (neurology)
5. Grave's disease (endocrinology)
6. rheumatoid arthritis (rheumatology)
7. bullous pemphigoid (dermatology)
8. scleroderma (dermatology)
9. ulcerative colitis (gastroenterology)
10. pernicious anemia (hematology)
11. Addison's disease (endocrinology)

12. Goodpasture's syndrome (nephrology)
13. primary biliary cirrhosis (gastroenterology)

The list could be expanded to greater length, but my argumentative point should be clear from the listed examples: Given that the theory of autoimmune phenomena must account for physiological disease processes in so many different kinds of organs and tissues, under so many different kinds of local conditions, the theory of autoimmune diseases is bound to be more complicated, less organized (at least at first), and generally less well systematized, than the various rather narrowly focused specialties noted in parentheses. The point I'm making against the Kemeny and Oppenheim model of reduction can be put another way. It isn't intratheoretic systematization that matters to us, or to the reductively minded practitioner, it is intertheoretic unification that matters. We don't care so much whether a theory is internally sloppy and disorganized as much as we care whether that theory has the potential to unifyto reduce to the same collection of terms and concepts phenomena from a number of domains of inquiry and practice previously thought to be distinct and unrelated to each other. Practitioners will pursue a reductive research strategy even one with a horribly unsystematized theory as the reductive base if they have grounds for believing that the unsystematized base theory is nevertheless the one dealing with the real causal and ontological aspects of the phenomena at issue. For example, consider the continuing research into neurophysiology as a reductive basis for all psychiatric phenomena despite the notoriously unsystematized, dare I say entirely conjectural, state of much neurophysiological theorizing at the present time. The gargantuan problems associated with simply localizing psychological functions in the nervous system may set the same sort of upper limit on being well systematized that potentially plagues quantum physics.

In summary, what really motivate the reductively minded practitioner are causal and ontological issues, not issues of explanation and systematization. Kemeny and Oppenheim were still operating within a positivist mind-set, one that sought to eliminate ontological motivations from science and replace them with epistemological substitutes. Instead of having reductively minded practitioners care about what causes what out in the real world, Kemeny and Oppenheim have them care mainly about how tidy and spiffy their theories are. But the large majority of scientists do care, and care very much, about what is going on in the universe or at least in the corner of it that they investigate. They are the philosophical and temperamental descendants of Thales, Anaximander, Anaximenes, and other Milesians of 2,500 years ago. Like Thales, Anaximander, and Anaximenes, most scientists are ontologists at heart. They wish to understand the cosmos, and to understand it as a cosmos, and that means that they wish to understand how its parts are unified into a cosmos. This presumption that the universe is a unified system of domains that form one hierarchical network is called the Unity of Science. A closer look at the Unity of Science research program will help us to understand why reductionism, despite its recent decrease in popularity, is a permanent feature of the landscape within philosophy of science.

Unity of Science Program

The Unity of Science was a concept developed within the positivist tradition in philosophy of science. The esteemed American philosopher Hilary Putnam and, once again, Paul Oppenheim, coauthored an influential study of the Unity of Science program in 1958 (although now Putnam would surely disown the notion). The idea at the core of the Unity of Science is deceptively simple: There are no special sciences, where by 'special' one means a science whose fundamental concepts are somehow incapable of being connected with or related to the fundamental concepts of at least one other science. For if there was such a special science, then presumably the phenomena it investigated would also be phenomena somehow determinatively unconnectible to any other kind of phenomena in the universe. The universe in such a case would contain an unexplainable discontinuity in it. It would fail to form one continuous fabric "all the way down" from the largest supergalactic superstring to the smallest subatomic particle. The universe in that sense would fail to be a cosmos. But how could that be? How could it be that certain theories from different domains "line up" in reductive order, but that this is merely a local achievement that there are theories that must remain outside the reductive network no matter how far science advances? Such a result is metaphysically obnoxious, if not downright bizarre. The bizarreness of special sciences can be brought out clearly if we read the lessons of current astrophysical cosmology correctly.

Cosmology is the branch of astrophysics responsible for giving an account of the current overall structure of the universe and how it got that way up to the present. Such a task requires tracing the evolution of the universe from its beginning, and that tracing involves understanding the best available theory of the origin of the universe the misleadingly named big bang theory. If the universe began from a singularity that "exploded," as the big bang claims it did, then the universe is bound to be a completely homogeneous entity; that is, there would be no ontological grounds for some parts of the universe being so qualitatively distinct that they can't be fit in theoretically with any other parts of the universe. To use a crude analogy, if a cake starts out as flour, eggs, butter, milk, chocolate, and salt, then how could what comes out of the oven as the finished product be something with certain properties in one part of it that are wholly incompatible-logically, causally, and theoretically incompatible with the properties in any other part of it? After all, its all just so much flour, eggs, butter, milk, chocolate, and salt, which have been mixed together and heated.

To return to the main path of our argument, a special science would be a science of special phenomena, phenomena that destroyed the homogeneity of the universe in virtue of their specialness. The positivist philosopher Herbert Feigl once called the theoretical laws about such special phenomena, the laws that would presumably be at the core of any special science that investigated such phenomena, nomological danglers. 'Nomos' is the Greek word for 'law'. Such laws would "dangle" outside the rest of the scientific network of theories, and therefore they are to be rejected for three reasons. First, they violate Ockham's Razor. Recall that Ockham's Razor requires that we do not multiply distinct entities and properties beyond the minimum necessary to explain the phenomena in question. Second, nomological danglers are inconsistent with the continuity of nature, a fundamental principle most scientists embrace. Why do they embrace it? Because of the third reason to reject nomological danglers: They are conceptually incompatible with the big bang cosmology, a cosmology that entails the continuity of nature.

Putnam and Oppenheim in their influential paper presented an argument for the Unity of Science that involved mereological premises. Mereology is the study of part/whole relationships. The two premises in question they called the principle of evolution and the principle of ontogenesis. The first principle holds that the universe evolved from smaller to larger in terms of organization; so, for any level of organization in nature, there was a time in the evolution of the universe when there were systems of that level but no systems of any higher levels of organization. The second principle holds that for any particular system at a given level of organization, there was a time in the evolution of the universe when the system did not exist but some of its parts on the next lowest level of organization did. Both these mereological premises suffer from serious problems. The principle of evolution assumes that there is a natural and unique way to decompose the universe into "levels" of decreasing organization, that there is one correct way to do this. However, there are different ways of cutting up a system into parts. It all depends on what your purposes are in decomposing the system. The principle of ontogenesis assumes something even more questionable: that it makes sense to speak of a system's parts existing before the system itself exists. But there are notorious identity problems with this way of proceeding. It would seem to be a safe bet that a system must either exist or have existed in the past for us to speak of some other thing being a part of that system. In other words, the prior existence of the system is a necessary condition for the existence of any of the system's parts. The property of being-something-which-will-be-a-part-of system S is not the same as the property of being-a-part-of system S. If this is so, then the principle of ontogenesis suffers from a fatal internal incoherence.

I do not think that the proponents of the Unity of Science program needed to base it on any mereological principles like those Putnam and Oppenheim offered in 1958. The mereological principles have a contrived air about them, a whiff of neopositivist desperation. The real motivations for belief in the Unity of Science research program have always been twofold: past reductive successes, and the continuity of nature as an implication of the big bang cosmology. These were the real epistemological grounds for the Unity of Science, not a series of arcane principles about parts and wholes. It will help to have a pictorial representation of the Unity of Science to understand the main critique of it that many philosophers of science believe killed it off for good. Even if those philosophers of science are wrong, and the Unity of Science is not dead, it still had to be modified enormously to defend it adequately against the critique. That modification made the Unity of Science nearly unrecognizable with respect to its original form. The modification changed it to a "nonreductive" Unity of Science.

To facilitate the coming critique of the Unity of Science I have split the higher-level sciences into the two classes depicted. The idea is that both physiology and geology reduce to different subbranches of chemistry. Note the converse directions of determination (for example, causation) and explanation. Determination moves from lower to higher and explanation moves from higher to lower. Crudely, parts determine wholes, and wholes are explained by their parts. Whether the ancient Pythagorean philosophers were correct and the universe is ultimately a purely mathematical entity is a subject of current dispute, so I have put a question mark next to 'mathematics' in the core sciences box. Certainly one can understand why the Unity of Science program as set out might be attractive to a philosopher who is sympathetic to science. The notion of a universe that is a cosmos is encapsulated in this picture. The various strands of science all fit into one fabric, and their specific systems of concepts boil down to one architectonic scheme of categories. There are no special sciences. Unfortunately, as we are about to see, the picture is an over simple distortion of what is really a much sloppier and murkier reality.

Putnam/Fodor Multiple Realization Argument against the Unity of Science

A sly reader will have noted that our discussion of the Unity of Science research program remained at a disturbingly non detailed level of description. It was all a bit too programmatic, with too many broad strokes and not enough narrow ones. How precisely would psychology connect up reductively with neurophysiology? How exactly does oncology (the medical science that studies cancer) reduce to immunology? Well, says the partisan Unity of Science supporter, those are mere details to be left to the grunt practitioners down in the trenches of research to figure out a likely story, replies the Unity of Science skeptic, if ever there was one.

Putnam himself came to reject the Unity of Science Program only a few years after he'd written with Oppenheim the classic presentation and defense of it. Putnam made his rejection of the Unity of Science turn on the alleged non reducibility of concepts within theories. The philosopher of psychology Jerry Fodor extended Putnam's point and turned it into a critique of the Unity of Science in which the incompatibility that destroys the possibility of reduction is an incompatibility between the actual phenomena the theories are about. The position that Fodor and Putnam both argued for in this way is called antireductionism. If Fodor is correct, then psychology is a special science in the strongest sense possible. What Fodor argued was that there exists a particular level of organization in nature at which a discontinuity occurs between phenomena of different types; so there is a theoretical discontinuity between the sciences that supposedly meet in the reductive hierarchy at that level of organization. Where is this disruption in the fabric of nature? It is where psychological phenomena meet neuro physiological phenomena. So, the theoretical discontinuity, the place where reduction hits a dead end long before it should, is where psychology meets neurophysiology.

To oversimplify Fodor's charge to the point of near vulgarity, the mental ain't merely the physical. To be sure, Fodor hedges and qualifies his claim (using what strikes some philosophers of science as a truly hair-splitting distinction without any real force), but in a way that seems to strip it of any interesting power and which it would not profit us to consider in greater detail here.

Fodor's argument depends on the viability of a very controversial philosophical concept: natural kinds. The concept of a natural kind comes from the ancient Greek philosophers, especially from the works of Aristotle, the first systematic biologist in the history of science. Aristotle claimed that science seeks to uncover not merely the accidental aspects of the phenomena it studies, but its essential aspects as well. For an Aristotelian biologist, the theory that investigates carnivores, for example, must uncover the essence of being a carnivore. A set of entities capable of having an essence forms a kind; and, if the entities in question are naturally occurring entities, the kind is a natural kind. All carnivores, to use our example, must share some essential property or properties in virtue of which they are all members of the natural kind denoted by our term 'carnivore'. What might such a kind-forming essence be like? Well, how about this: Exclusively meat-eating is the essence of a carnivore.

Hence, carnivores are a natural kind, and they therefore figure in the causal network of the biological world. For example, carnivores causally interact with noncarnivores of various natural kinds: herbivores and omnivores, for instance. Furthermore, carnivores themselves come in various natural kinds: lion, hyena, jackal, cheetah, and so on. Each of these natural kinds has its own kind determining essence. In physics there are natural kinds: electron, proton, neutron, lepton, meson, quark, and many others. Within immunology there are natural kinds. NK cells form a natural kind, one whose essence might be given as, non-MHC restricted, nonsensitized killer of altered-cells (the reader need not worry for now about the technical modifying expressions take my word for it, the essence of an NK cell, if it has one, is something close to this).

What natural kinds allow a practitioner to do, among other things, is to cut the domain of inquiry along its natural "seams." For example, no biologists practice their science as though there exist objects of investigation within their chosen domain of inquiry that are called "zebons," where a zebon is a mammal composed of the front half of a zebra and the rear half of a lion. There simply is no such kind of critter in nature. Notice certain epistemological and methodological implications that flow from the absence of a natural kind: There are no theoretical laws about zebons, and so there can be no predictions about them. Likewise, zebons do not participate in any causal relationships with other natural kinds that actually exist; so, zebons cannot figure in any explanatory relations involving natural kinds that actually exist.

That the universe presents itself to us with natural kinds already in it is a central presumption many scientists make, perhaps most of them unconsciously so. That a theory's fundamental laws should be about the natural kinds that occur in that theory's domain is another central ideal at which many scientists aim, again perhaps without an explicit awareness on the part of many of them that they do so. Jerry Fodor certainly presumed that psychologists are among such scientists. Like any other domain of inquiry, psychology seeks to uncover the genuine causal relations between the genuine natural kinds that form the seams of the psychological level of phenomena within the universe. Psychological laws of nature will relate natural kinds to each other, natural-kind properties to each other, and natural-kind events to each other. But that means, argued Fodor, that psychology cannot be reduced to any theory or collection of theories from any sciences below it in the hierarchy presumed by the Unity of Science; for when we understand how it is that psychological states and properties of organisms are physiologically realized in the nervous systems of those organisms, we shall see that the terms constructed in the reducing theory to form the bridge laws, in many cases, will not denote natural kinds.

Why not? Because a given psychological state or property is "multiply realizable" in the physiology of a typical nervous system. Take the psychological state of believing that there is life after death. There will be no one, single, physiological way for that state to be realized in the neurophysiology of a human being. Suppose Casper and Daphne both believe that there is life after death. There may be nothing physiologically similar about the way that mental state is represented in the nervous systems of Casper and Daphne. Yet, in each of their nervous systems, there is some neurological representation of a mental state with that specific content: believing that there is life after death. Suppose we begin to survey the nervous systems of people who each believe that there is life after death. We may find that out of, say, 40,000 such people, there are a total of 28,374 different ways that particular mental state can be neurophysiologically realized in the human nervous system. But this destroys the possibility of constructing a bridge law linking the mental predicate 'believes that there is life after death' with a co extensive predicate in neurophysiological theory, claims Fodor; for the half of the bridge law representing the neurophysiological term would be an artificial disjunction a huge predicate composed of separate neurophysiological descriptions disjoined to each other (linked by 'or') that does not denote a natural kind. But laws of science link natural kinds, not artificial kinds, so the bridge law would not be a law at all, but rather some monstrously complicated and predictively useless correlation between a natural kind mental state and what should we call it, a term that refers to no piece of nature?

Let 'M' denote the mental property of believing that there is life after death. Let 'NP_1', 'NP_2', 'NP_3', and so on denote each of the 28,374 different neurophysiological realizations of the psychological property of believing that there is life after death. Then the bridge "law" needed to link this mental state with the neurophysiological realm would look like this:

for all organisms o, o has M if (and possibly only if) (o has NP_1 *or o has* NP_2 *or o has* NP_3 *or o has* NP_4 *or o has* NP_5 *or o has* NP_6 *or has . . . or o has* $NP_{28,374}$*)*

The last half of this statement does not pick out a natural piece of the universe. Physiological state NP_{64} may be nothing like physiological state NP_{567}, which in its turn may be nothing like physiological state $NP_{7,441}$, and so on. The individual states connected by the or's have no one thing in common. Hence, the entire last half of this supposed bridge law is a specially concocted disjunctive term that only exists because some crazy reductionist wants to reduce the psychological level of nature to the neurophysiological level of nature. The statement is predictively useless; for we would not know which of the 28,374 disjuncts was true of any given person who believes that there is life after death. All we could say of any given case is that some disjunct or other must hold good. But that is so predictively useless as to be a meaningless exercise. But if the statement is predictively useless, it is also explanatorily useless.

The reader may wonder why there are so many different ways in which the same mental state could be realized within the nervous system of a human being. There are good grounds for thinking that this is so, but they involve some rather technical distinctions and arguments within the philosophy of mind, distinctions and arguments not directly relevant to our purposes in this text. Perhaps it is sufficient here simply to note that the content of many mental states is characterized functionally; that is, their content is characterized causally. Mental state M is the mental state of believing that p, or wishing that p, or whatever, because it is that state which is caused by certain input stimuli and that in turn causes the organism to produce certain output behavior. If this is correct, then the content of a mental state has nothing to do with the particular kind of physical material in which it is represented. If being angry at the current U.S. president is the mental state with that contentit is anger (not joy or indifference) directed at the current U.S. president (not your brother, not a past U.S. president)in virtue of being that state of the organism caused by input experiences el through e_n (for some number n greater than 1) and which in turn causes the organism to produce output behavior b_1 through b_n, then any organism for which e_1 through e_n is connected with b_1 through b_n in this way is an organism angry at the current U.S. president, no matter what that organism is made of, no matter what its internal structure is. If space aliens land in a spaceship tomorrow and it turns out that they too are angry at the current U.S. president, that is perfectly possible even if the space aliens turn out to have a physiology of boron, nickel, tin, and silicon (instead of a human hydrocarbon physiology).

Suppose you and I are both thinking of the first time human beings landed on the moon. This is a mental state functionally characterized. That means that there doesn't have to be anything, not one single thing, similar in your brain and my brain in virtue of which we are both in this same mental state. That is why Fodor's argument against reductionism gets off the ground: The multiple physical realizability of mental states will result in bridge laws that aren't laws at all because their physiological sides do not pick out natural kind properties, objects, or events in the physiological domain of nature.

The Putnam/ Fodor multiple realization argument created much consternation in philosophical circles. Some reductionists tried to duck its impact with a "so what?" response. Other philosophers jeered at the argument's use of the concept of natural kinds on the grounds that there are no such things; or worse, on the grounds that there are natural kinds but we get to say what they are, not nature (and so, the last half of such a bridge law is as natural as you want it to be). Still other philosophers took the argument to heart and said good-bye to reductionism. There is at least one special science after allpsychology.

To most philosophers none of the above reactions seemed adequate. The argument must be given its due respect, but neither must the baby be thrown out with the bathwater. One such thinker was the philosopher Jaegwon Kim. Kim sought to recognize the multiple realizability of mental states while at the same time he wanted to rescue the ontological point of reduction. The main philosophical purpose of reduction is to achieve an ontological economy of sorts, to show that there are not two ontologically distinct kinds of phenomena going on, but only one kind of phenomena going on. The mental just is the neurophysiological, at least it is for human beings. Kim achieved his reconciliation by revamping a heretofore ignored kind of asymmetric covariance among distinct families of properties, an asymmetric covariance that is due to the fact that one family of properties is determinatively dependent on a qualitatively distinct family of properties without properties of the former family necessarily being reducible to properties of the latter family. Further, the determinative dependence at issue is not necessarily ordinary causation; for ordinary causation is law like and hence holds only between natural kinds (and therefore would support a reductive relationship between cause and effect). This interesting brand of asymmetric dependence Kim called supervenience, and to its consideration we now turn.

Supervenience to the Rescue

When the broad-spectrum antibiotic penicillin was first approved for use by physicians it was prescribed as though it was a wonder drug. Doctors used it to treat every minor sore throat and puffy big toe. Medical microbiology was uncertain at the time precisely how an antibiotic kills its target microorganisms and that ignorance of detail was filled in with an unsupported impression that penicillin killed practically anything small enough to require a microscope to identify. The resulting over prescription produced, among other results, a host of penicillin-resistant strains of microorganisms that otherwise would not have evolved (or, at least, they would not have evolved as soon as they did). It is perhaps in the nature of scientists to seize on a new instrument, material, or concept in the hope that it is a universal tool capable of solving the most severe theoretical, experimental, and practical problems that hound daily practice in their fields.

Philosophers are only slightly less trendy in the practice of their own chosen field. New concepts and arguments are often seized on by philosophers in the hope that they are tools that at last will settle some long-standing dispute or render clear some cloudy issue of perennial endurance. Supervenience has suffered just such a fate. It became the "lab" concept in metaphysics and philosophy of science in the years after Kim introduced a precise model of it. But just like in the case of penicillin, its popularity rode a wave of vagueness, imprecision, and sometimes outright distortion. Many philosophers helped themselves to supervenience without first taking the time either to construct a precise model of it or to learn such a model invented by someone else. What supervenience is to one philosopher is often not the same as what it is to the philosopher up the road a piece. As a result there has been a proliferation of different concepts all claiming to be concepts of supervenience. This situation is not unlike a case in which fifty different psychiatrists present fifty different conceptions of what psychoneurosis is, or what schizophrenia is. Kim himself has noted the lack of a prior ordinary-language use for the terms 'supervene', 'supervenient', and 'supervenience', and that in such a situation these technical terms mean pretty much what the philosophical specialist wants them to mean. In the absence of corrective pressures emanating from ordinary speech, the main constraint on our use of a specialized term of art is its degree of pragmatic usefulness "under fire." What are we to do in the face of this profusion of models of supervenience? The present author was a student of Kim's, and he believes that Kim's original model of supervenience is the most useful and conceptually richest model of the lot of them. Accordingly, in this text we will understand supervenience as it is characterized in Kim's model of it, to which we now turn.

Kim's Model of Supervenience

The original purpose behind supervenience was a metaphysically deliberate one: to construct a concept of asymmetric dependence under which the determinative relation represented by the concept in question is: (1) not necessarily causal (it does not require law like regularities); (2) nonreductive (to allow for the manyto-one aspect of multiple realization); and (3) ontologically significant (strong enough to support, say, a general scientific physicalism, without being so strong as to be reductive). In short, what supervenience was originally designed to save from the aftereffects of the multiple realization argument was noncausal, nonreductive physicalism. 'Physicalism' denotes what used to be called materialism, the view that the universe is ultimately an entirely physical system. What physicalism implies is that a mild version of the Unity of Science holds good: Ultimately there are no phenomena in the universe which cannot be understood in terms of the concepts of physics. So, in a way, supervenience was called on to salvage a toned down version of the Unity of Science, a version in which the respective sciences line up in order of determinative dependency without lining up in reductive order.

The first tricky aspect of supervenience is that it is a relation that holds between natural groupings or what Kim called families of properties. One way of understanding this is to note that competent supervenience theorists rarely talk about a particular property supervening on another particular property (except as a limiting case for illustrative purposes only). They always talk in more general terms such as, "the mental supervenes on the neurophysiological" or "allergic phenomena supervene on immunological phenomena." The former statement can be interpreted along the lines in which the latter is expressed: mental properties and events (as a family of such) supervene on neurophysiological properties and events (as a family of such). Kim constructed his model of supervenience originally for families of monadic properties (that is, properties that are not many place relations). Let the family of psychological properties attributable to higher organisms be represented by $M^?$. The point of the '?' after the 'M' is to indicate that we include in $M^?$ properties formable by applying simple logical operations to the basic properties of $M^?$, operations such as conjunction, disjunction, and negation. Let the family of neurophysiological properties attributable to higher organisms be represented by $N^?$ (where '?' again indicates simple compounding operations to make more complicated properties out of simpler ones). Then the psychological supervenes on the neurophysiological if and only if the following holds:

> *necessarily, for any two objects x and y, if x and y have exactly the same $N^?$ properties, then they have exactly the same $M^?$ properties; and, if x and y differ in $M^?$ properties, then they differ in $N^?$ properties.*

Fans of metaphysics should note that supervenience comes in different strengths-three, to be exact: global, weak, and strong. The version above is of middle strength, called weak supervenience. Arguments can be heard that weak supervenience is not robust enough to qualify as an interesting resolution of either the traditional mind-body problem, or the Putnam/ Fodor multiple realization argument. I would prefer to use strong supervenience, but its use here would require the introduction of possible-world semantics into the discussion, a highly technical and complicated detour we cannot afford the chapter space to pay for.

Supervenience as applied to the case of psychology comes to this: There is no difference in psychological properties without a difference in neurophysiological properties, but there can be a difference in neurophysiological properties without there being a difference in psychological properties. This last remark should be very significant to the reader. It indicates that the psychological can supervene on the neurophysiological in a one-to-many fashion: The same psychological state can be neurophysiologically realized in many distinct ways. What cannot happen if the psychological supervenes on the neurophysiological is that the same neurophysiological state can be realized as more than one distinct psychological state. The neurophysiological varies if the psychological varies, but the psychological does not necessarily vary if the neurophysiological varies. The determinative dependence is asymmetric. The core question then becomes: Is asymmetric covanance a relation strong enough by itself to sustain a watereddown Unity of Science program?

Modest Physicalism

Kim has recently suggested a negative answer to the question that closed the last section. The results of work by many supervenience theorists in recent years indicate that asymmetric covariance is logically independent of asymmetric determinative dependence: Two families of properties can stand in a relation of supervenient asymmetric covanance, and it can be the case that neither is determinatively dependent on the other. But determinative dependence would surely be one minimum requirement of any Unity of Science program. If the various sciences cannot be lined up in reductive order, then at least they should line up in order of determinative dependencethey should, that is, if the Unity of Science is to have even the slightest interest. If Kim is correct, then it is a separate task to show that a form of determinative dependence underlies a given supervenience relation. Simply pointing out that supervenience holds will not be enough.

Suppose that allergic properties supervene on immunological properties. This is a supposition with much evidence behind it. Suppose further that immunological properties supervene on genetic properties, for which again, there is much evidence in support. Suppose further that genetic properties supervene on physicochemical properties, a supposition that has by now become part of standard scientific knowledge of the biological world. Once we have hit the physicochemical domain, the rest of the determinative order is pretty much set all the way to fundamental particle physics.

Kim has argued that any supervenience relation worth the name is transitive. If A supervenes on B and B supervenes on C, then A supervenes on C. The presence of an intervening third family of properties in the supervenience chain from A to C may serve to illuminate the determinative ground of the supervenience relation of A on C. If this is so, then we at least have rendered plausible a "modest" version of physicalistic Unity of Science. But we need to note one important caveat. Supervenience, in Kim's view, is not correctly thought of as a self-contained "kind" of determination at all. It is rather a pattern of asymmetric covariance, and we may legitimately ask of any such relation of asymmetric covariance: What is the determinative ground of the supervenience relation? The idea is that in different domains of inquiry, a different kind of determination may ground, account for, the supervenience relations that hold in that domain. In chemistry and physics, for example, the supervenience of the chemical on the physical is determinatively grounded by mereological relations: The properties of larger wholes are determined by the properties of their constituent parts. The chemical properties of molecules, for example, are determined by the physical properties of their atomic constituents. Now in cognitive psychology and neurophysiology, it is presumably an open question subject to controversy just what kind of determinative relation grounds the supervenience of the psychological on the neurophysiological. Since the advent of the multiple realization argument, most theorists working in this area grant that the determinative relation cannot be causal (for there can't be the kinds of laws liking natural kinds required for causation). But what the determinative relation is in a positive sense is under dispute. The behaviorist, for example, might argue that the determinative relation in question is semantic: Psychological states of organisms supervene on physically characterized behavioral states of such organisms because of certain "analytic" semantic connections that hold between the linguistic expressions that denote psychological states and the linguistic expressions that denote physically characterized behavioral states. The supervenience is determinatively grounded in relations of semantic meaning. The functionalist psychologist would, of course, take a slightly different line of approach.

What the demand that we distinguish supervenience from its ground amounts to is an admission that supervenience does not itself constitute an explanatory basis for anything. The existence of a supervenience relation between property family A and property family B suggests that there is a determinative priority between the two families of properties; but it does not indicate the specific nature of that determinative relationship (is it mereological, causal, semantic, or some other kind of determination?). And because it does not indicate the specific nature of the determinative relationship, supervenience does not by itself count as an explanatory relation between the two families of properties. Supervenience suggests that there is some determinative relation between the property families in question that is sufficient to underwrite an explanatory relation between them; but the kind of determination may be different in different domains. Does it follow from this that the kinds of explanatory relations that obtain between different phenomena will also vary from domain to domain? Is what counts as a scientific explanation radically different in different domains? Or is there a single unified model of what is involved in explaining scientifically a particular phenomenon? The traditional view, the one that flowed naturally from the positivist model of science, gave an affirmative answer to the last question and a negative one to the others. Nowadays, however, there is less acceptance on the part of philosophers of science of such a monolithic conception of scientific explanation.

15

Constructivist Challenges in Science

Introduction

The previous consensus view that science, at least on occasion, discovered a preexisting objective reality gave way to alternative critical accounts of science that stripped it of any such achievement. These new critical accounts were invariably relativist in epistemology, and, for the most part, antirealist in ontology but the relativism and antirealism in question were not the traditional philosophical brands of relativism and antirealism. The new forms of relativism and antirealism were born within the nest of the social sciences. Many Kuhn-inspired critics of science took Kuhn's ultimate point to be that the philosophy of science, as traditionally conceived since the time of Immanuel Kant, was dead. Its old job would be taken over by a successor discipline called variously the sociology of science, the sociology of knowledge, or science and technology studies. And this new discipline would not go about the old job in the old way. One leading spokesperson of this new discipline is the sociologist David Bloor, who published a treatise outlining the details of the new successor discipline to philosophy of science. Bloor called this new successor discipline the sociology of knowledge, and the particular type of it that he advocated he called the Strong Programme.

Strong Programme

The Strong Programme was born partly out of a reaction to traditional philosophers of science who sought a compromise between Kuhn and traditional neopositivist philosophies of science. With Kuhn, these philosophers agreed (rather grudgingly) that different paradigms are incommensurable. But, unlike Kuhn, they argued that such incommensurability did not cut us off from giving a *rational* assessment of competing paradigms: for it is possible to judge paradigms by how well or poorly they solve the research puzzles they set for themselves, in their own terms, you might say. The upshot of this compromise model of science was that *non*rational "social" factors in the practice of science were to be called on *only* to explain failure and error within a paradigm-based research tradition. Wherever a paradigm-based form of inquiry met with success, however, the explanation for that success was that the paradigm was accurate, that the model of reality it contained was true. Thus, the compromise position amounted to the claim that nonrational social causes are to be appealed to only when explaining scientific error, never in explaining scientific success, thereby securing the essential rationality of science even in the face of major error.

This asymmetry is unacceptable to Bloor and his fellow sociologists of knowledge. Against it, the Strong Programme holds that false beliefs and true beliefs in science are to be explained by the same kinds of social, nonrational causes. That is, in the Strong Programme one is never allowed to explain why practitioners of a science hold a particular belief by saying "because that belief is true." This is the source of the name "Strong Programme"; for the "weak" programme in sociology of science is the one advocated in the compromise view, in which only erroneous beliefs receive explanations in terms of nonrational "interference" factors. On the contrary, Bloor insists, being true of reality does not explain why a true belief comes to be held by scientists. The usual reply by philosophers of science is that, if truth does not serve, in any capacity whatsoever, as a constraint on theoretical belief in science, then the content of science would surely float free of any theory-independent reality "out there" so free of it that the content of science in the end wouldn't be *about* that theory-independent reality at all, but about something else. Hence, Bloor's Strong Programme cannot avoid either antirealism or an "anything goes" form of relativism.

Bloor is aware of the above charge and he replies to it that the Strong Programme is not necessarily committed to antirealism; nor, for that matter, is it necessarily committed to an "anything goes" form of relativism. He does admit that the Strong Programme is a relativist model of science, but it's a brand of relativism of a rather modest sort (that is, less than an "anything goes" relativism). These are claims that a philosopher of science ought to be duly suspicious of, and we shall be very insistent ourselves that Bloor make good on them. The reason for our suspicion is that the Strong Programme, as set out by Bloor, aims for the outcome of showing that *scientific knowledge is socially constructed rather than discovered;* and, any time somebody aims for that outcome, the worry arises that antirealism and an "anything goes" version of relativism are going to crash the party urinvited despite the host's best efforts to bar the door against them. Bloor claims that the Strong Programme is not committed to antirealism because a sociologist of knowledge working in accord with the Strong Programme certainly admits the existence of a practice-independent and theory-independent material world. No advocate of the Strong Programme, says Bloor, is an idealist. What the Strong Programme asserts is that the nature and structure of the theory independent material world is so filtered through "socially sanctioned metaphors" that its nature and structure do not significantly effect the content of scientific theories. The idea seems to be that the practice-independent world is a kind of unknowable something-or-other we are cut off from any direct contact with; hence, science is not about this world, it is about this world under descriptions and explanations licensed by contingent and conventional social metaphors and relationships. It follows that the reality that science reveals is not the world's reality, it is the reality of our social forms of life. In the years since Bloor's manifesto appeared, sociologists of knowledge have done a great deal of work producing case studies designed to show how the content of even true beliefs in science is determined, not by evidence regarding the structure of the independent reality under investigation, but by nonrational social factors such as the political commitments of the practitioners in question, the unconscious projection of social relationships onto nature, desire for fame, etc. This sets up an argument that Bloor does not actually give, but which I claim is a fair representation of the Strong Programme:

1. True beliefs in science are just as caused by the interpretation of data via projected social metaphors as are false beliefs

   therefore

2. The content of scientific beliefs is as socially determined as anything else, which is to say it is more socially determined than it is non socially determined (the contrast here is with determination by what is the case about a theory-independent reality)

   therefore

3. Scientific knowledge is socially constructed rather than discovered.

   The central problem with the Strong Programme is that Bloor, despite his best efforts to the contrary, cannot prevent either antirealism or an "anything goes" form of relativism from crashing the party. To see why, let us take realism to be indicated in the last remark of the following conversation:

4. Q: Why does scientist S hold belief B?
5. A: Because holding B meets with success in Practice.
6. Q: Why does holding B meet with success in practice?
7. A: Because B is true (of a theory-independent reality).

   Answer 7 is the realist answer to question 6. The Bloorian social constructivist, however, is prevented from giving answer 7 by the rule in the Strong Programme that truth is never an explanation for why a belief is held. On the contrary, the Strong Programme must substitute a different answer to question 6:

8. Because current social metaphors license a verdict of success with respect to using belief B in practice.

   But 7' renders the Strong Programme vulnerable, I claim, to an "anything goes" form of relativism by the following argument:

9. Success verdicts are licensed by particular social metaphors and relationships that are projected onto nature in scientific method and theory.
10. There are no objectively based limits on which social metaphors and relationships could come to dominate a culture.

therefore

11. In principle, anything could come to go in science.

Bloor would surely object to the conclusion directly on the grounds that science must meet with some minimum level of success at manipulating the material world per se; hence, it is just plain false that anything could come to go in science. But the previous argument (4-7), I claim, cuts off this response, for that response helps itself to something Bloor is not entitled to given his other claims: a social-metaphor-independent notion of success at manipulating the material, theory-independent world.

I am arguing that the Strong Programme is not entitled to use the idea of success at interacting with a practice-independent reality to escape "anything goes" relativism because such success will be just one more socially determined and heavily filtered judgment, not about reality, but about reality under socially sanctioned description. The problem, as I see it, is the same one we've had now for almost 200 years since Kant: If one is serious about the Kantian distinction between phenomena and noumena, between things-as-interacted-with-via-social-metaphors and things-in-themselves, then, in the end, one cannot escape a completely skeptical antirealism about things-in-themselves, or complete relativism concerning our theories of things-in-themselves. If you conceive of the theory-independent world as a remote and unknowable thing-in-itself, then of course you will be led to the conclusion that we can know nothing of it, and its nature and structure therefore cannot significantly constrain the content of our knowledge, scientific or otherwise. Bloor defines true beliefs to be those that work without realizing that to escape antirealist relativism he shall need a social metaphor-independent notion of "working." But such a notion is not available to him, for verdicts about which beliefs work and which ones don't work are as much under the sway of social metaphors that float free of the theory-independent reality as anything else.

We should note the likely fact that Bloor would object strongly to premise 9. Bloor's view is that the Strong Programme is the one model of science that can honor science's objectivity, because it locates that objectivity in the proper place: the social. Bloor holds that *the objective is the social,* and it couldn't be anything else, because the theory-independent world is an unknowable thing-in-itself; hence, the only source of inter subjective agreement and theoretical stability is society itself. This is a fuzzy argument, and one senses that the term 'social' is being so stretched out to cover so many different kinds of phenomena as to bleed it of all significance. Bloor's strategy is to claim that only the social world could provide the kind of stability over time that underwrites intersubjective agreement, and only intersubjective agreement could underwrite objectivity. Premise 9 is false, Bloor would claim, because no one person or group of persons can alter at will the social metaphors and relationships that form the structural components of society. Not just any social metaphor could be set up to license success verdicts, seems to be Bloor 's suggestion. Society is bigger than any elements within it, and it has a dynamic of its own independent of the personal psychological desires of its members. Will this response by Bloor work? It doesn't seem very likely that it will. Nothing in the case studies he presents suggests that there are fixed limits on the different forms social metaphors and relationships can take in human life, on which ones can come to dominate a culture. For example, one might have thought that reason itself would supply fixed limits that the fundamental principles of logic and mathematics would simply rule out certain forms of social metaphor and practice on the grounds that those social metaphors and practices are incoherent, inconsistent, or absurd. Unfortunately for Bloor, he argues the very opposite. He insists that not even logic and mathematics contain fixed truths that could supply fixed limits to the forms of social evolution among human beings. A closer look at what he says about this particular issue is instructive.

The central claim in the Strong Programme is that the content of scientific theories is not significantly affected by the nature and structure of the theory independent reality that is nevertheless, in some Kantian fashion, "out there." This claim must be filled in for specific cases of successful science. In other words, even in positive cases of success (rather than negative cases of error), the content of the theories in question is determined by nonrational social factors not by "the reality" the theories are incorrectly thought to be theories about. Bloor attempts to illustrate this central claim with a number of case studies from, surprisingly, the science of mathematics. Unfortunately, there is a tendency for his case studies not to show what he thinks they show. *Instead of illustrating that belief content is a function of projected social metaphors, what Bloor's case illustrations show is that the Quine-Duhem Thesis is alive and working well in science.* Bloor takes cases in which practitioners respond to an anomaly by adjusting other background beliefs rather than by adjusting the foreground test belief to show that it is not the alleged theory-independent reality under study that is constraining the adjustment, but rather social factors are the source of the adjustment. But these cases of adjustments made to background beliefs in order to deal with anomalies need not be taken to show any such thing; for they can be seen more accurately as cases in which the adjustments in question were *in part* driven by what is the case about a practice-independent reality under investigation. It is an open question what drives such adjustments. To take an example Bloor investigates extensively, the ancient Greeks took the "proof" that the square root of 2 cannot be represented as a fraction composed of integers to be a proof that the square root of 2 is not a number at all, whereas moderns take it as a proof that there are certain numbers that are numbers, but "irrational" numbers. The ancients held that there is only one kind of number and the square root of 2 isn't among that one kind. The moderns believe that there are multiple kinds of numbers and the square root of 2 is a member of one of the more exotic kinds. Rather than treating this case as an example of an adjustment to background beliefs that counts as progress in ferreting out the structure of a reality which is independent of mathematical inquiry that counts as a case in which moderns discovered something the ancients overlooked Bloor interprets it as showing that there is no social-metaphor-independent fact of the matter whether the square root of 2 is a number or is not a number. I put the word 'proof' in quotes a few sentences back because one lesson Bloor draws from this case is that the notion of proof itself is a social creation. Proofs, he says, have no intrinsic meanings. To call something a proof is a "moral" act, it is to place a social approval upon it and take it to establish a piece of knowledge. The intrinsic ambiguity of formal demonstrations renders the force of logic and reason entirely reactive to cultural and social interpretations. One problem here is that Bloor falls to appreciate the fact that, if he is right, it strengthens even more premise 9. If even formal proofs in logic and mathematics can be reinterpreted to mean something other than what they are thought to mean, then "anything really could come to go," at least in principle, for not even math and logic can serve as universal and absolute constraints on our belief formation or behavior. Another problem with Bloor's argument has a more systematic character. The case of mathematical proof might be rather atypical. Traditionally, philosophers have given mathematics a special epistemological status with respect to the rest of our empirical inquiry. A case could be made independently of specifically social constructivist assumptions that mathematics involves a large dose of conventionality that mathematical concepts are especially conventional to begin with a matter of how we agree to agree on the meanings of special symbols and procedures. If that is so, then Bloor has no business generalizing from the case of mathematics to other domains of inquiry where things may not be as convention based. It is difficult to imagine Bloor being able to make the same sort of argument about proof in, say, immunology, where there is presumably (or so says the realist) a wealth of non convention-based physical structure that impacts on what counts as conclusive evidence for what else.

Roll Over Socrates

The Strong Programme is beginning to look like a Trojan horse inside the walled City of Science. Bloor promises in his argument for the programme that it will neither damage, cheapen, nor insult science, but instead it will place science on the only proper footing on which its objectivity could be legitimately respected. What we have seen of the Strong Programme so far, however, suggests that it is basically Kantian idealism dressed up in the latest social science clothing, Kantian idealism remove from the individual mind (which is what Kant cared about) and placed in some sort of collective social mind.

The writings of the Strong Programme advocates are unfortunately filled with mishandled case studies such as the one above about the proof that the square root of 2 is a non repeating decimal number. It is perhaps not surprising that such cases are mishandled by proponents of the Strong Programme, for one notable feature of the sociology of science is the extent to which it depends on either redefined terms or vaguely defined terms that otherwise have a relatively clear sense within philosophical discourse. Early in his treatise, for example, Bloor attacks traditional epistemology as it is practiced by philosophers. He announces that the sociology of knowledge takes the term 'knowledge' to refer not to beliefs that are justified and true, or reliably produced and true, or any of the other characterizations currently offered by philosophers, but rather the term refers to whatever beliefs are "socially endorsed" by a given community. Truth is not a necessary condition of knowledge, Bloor boldly asserts. So, if the belief that the square root of 2 is not a number at all is a socially endorsed belief in a given culture, then the members of that culture can be said to know that the square root of 2 is not a number. This virtually assures relativism by definition, and Bloor defends this move by suggesting that the traditional definitions of knowledge in philosophical epistemology beg the question against relativism by assuring an absolutist form of knowledge. In fact, Bloor is refreshingly honest about this sort of polemical aspect to the sociology of knowledge. He admits that the sociology of knowledge, certainly as it is embodied in the Strong Programme, is a competitor discipline to philosophy of science and to philosophical epistemology. They compete on the same intellectual turf, and the sociology of science wins out over philosophy of science and philosophical epistemology, in Bloor's view. A philosopher is pretty much at a loss what to say about Bloor's redefinition of knowledge to mean something like popular belief. To amend a phrase from Chuck Berry's "Roll Over Beethoven," Bloor's message to philosophy might be said to be, "Roll over Socrates!" Any competent student of philosophy is well aware of how intensely and successfully that mighty stonecutter of ancient Athens argued for the distinction between knowledge and popular opinion.

It is clear from his writings that Bloor understands the Strong Programme in the sociology of knowledge to be a form of naturalism. As such, the programme is opposed to what Bloor considers the nonnaturalism of traditional philosophical epistemology. What Bloor wants to show is that knowledge, truth, and other epistemological concepts can be given perfectly natural, causal characterizations in terms of contingent, socially determined explanatory metaphors, and that therefore there is no need to posit an abstract "world" where eternally abiding logical entailments, relations, and conceptual truths "reside" and thereby fix in place an objective epistemological reality. Certainly he is correct that much of traditional philosophy has opted for such an otherworldly treatment of abstract concepts like truth, knowledge, and analytic entailment.

But not all philosophers stick with the traditional party line on these matters. Quine himself railed against thinking of meanings and concepts as quasi-objects in some heaven of abstract entities where they hover about with timelessly fixed attributes that our linguistic practices here on earth must respect by matching correctly. Bloor is a devout anti essentialist, in the jargon of contemporary philosophy, and therefore he seeks a model of knowledge and science in which neither involves the uncovering of the supposedly fixed essences of things. He believes that specifically philosophical accounts of these matters are hopelessly essentialist, therefore they are all bogus accounts. Philosophical accounts of concepts seek to mystify them, in Bloor 's view. He means by this that philosophical accounts seek to portray the concept under analysis to be one so mysteriously abstract and remote from ordinary causal and material life as to be inaccessible to anyone but the specially initiated professional philosopher. This will ensure, it seems, that the rest of the world will always need philosophers around to explain these mystifying notions only they understand. Professional philosophy is thus mostly stripped of its cognitive merit and debunked as being a form of self-serving intellectualism.

Against the mystification of knowledge by philosophy the sociology of knowledge supposedly stands as a bastion of intellectual egalitarianism, as the champion of the common person who lives in the everyday social world and fights the good fight against the goofy philosophers addicted to their otherworldly Realm of the Conceptual. If a social constructivist movie was made of human intellectual inquiry, philosophy would be like the proverbial "bad guy" of an American Western, the one with the skinny mustache, the squinty eyes, the bad skin, and the knurled hands. But ordinary folk terrified by the sour-souled philosophers should take heart. The sociology of knowledge has arrived in town for the showdown, and you can bet the house that it will be the evil mystifiers of human knowledge who shall eat the dust after the guns blaze at high noon. For example, according to Bloor, philosophers invented the notion of *necessity* so that they could succeed in mystifying concepts like causation and logical truth. In fact, argues Bloor, necessity is actually nothing more than a projection onto nature of a central social metaphor in all human societies, *moral authority*. To claim that such-and-such a statement is a logically necessary truth, or that such and-such a sequence of events was causally necessary, is merely to invest that sentence and that sequence of events with the sort of authority over human affairs that moral taboos and commands have. The laws of a science, for example, do not represent essential connections between structural aspects of some inquiry-independent reality, rather they separate out certain patterns within our practices in the area of inquiry in question and invest them with the power of moral authority. These are wonderfully clear examples of how the Strong Programme supposedly debunks and demystifies scientific concepts that philosophers of science allegedly insist on treating in some more mystifying and essentialist way.

One wants to ask the obvious question: If necessity is a projection onto nature of the social metaphor of moral authority, then because it is a mere *projection* by us, does this mean that there are no morally authoritative necessities "out there" in the inanimate material world? Bloor seems strangely cavalier about this possible implication of his argument. A projection is imposed from outside, and it can be a distortion of what it is imposed on or else it can fit it perfectly. In science, among other things, we want to know what must be the case about things, if anything must be the case about them. Apparently, we cannot know that, according to Bloor, for the most we can know is what the projected moral authority tells us we think must be the case about things, and then only in the sense of what we think *ought* to be the case about them. The reader would do well to file this result away in a safe place, for the feminist social constructivists whom we will investigate in the next chapter will seize on this aspect of social constructivism and make it one of the centerpieces of their critiques of science.

Case Method with a Vengeance

The sociology of knowledge is an empirical discipline. It will stand or fall on whether its pronouncements are consistent with the relevant data. The relevant data are case studies in the history and/ or practice of science. It is no wonder then that social constructivist models of science are heavily dependent perhaps even more so than Kuhn himself wason case studies of scientific practice and method. Bruno Latour and Steve Woolgar are two prominent social constructivists who are specialists in the right sort of case studies.

The idea is to do the case studies, not the way a historian would, but the way an anthropologist doing field work would. This is what Latour did when he showed up at a research laboratory housed in the Salk Institute in California, a laboratory specializing in the investigation of hormones that originate in the nervous system. He wanted to observe practitioners working in the lab in the hope that they might follow the social construction of a scientific "fact" from beginning to end. Woolgar later joined Latour to coauthor a study based on Latour's observations. They wrote with a heavy dose of the self-referential plural despite Woolgar's absence from the lab. My own remarks below will respect their preference in this matter by discussing their study as if it was based on a jointly experienced sociological field exercise.

Latour and Woolgar admit to approaching science with an attitude not unlike that with which field anthropologists approach the study of a primitive and alien culture. Science to them is alien in the same way a previously unknown tribal culture is alien. Further, just as no well-trained anthropologist would take the primitive natives' own interpretation of their culture as a correct account of it, so Latour and Woolgar argue that it is not appropriate to take scientists' own interpretations of their work as correct accounts of it. The sociologist of science must avoid "going native," as they put it. On the contrary, science is to be accorded initially no more rationality as a knowledge-gathering method than is, say, primitive witchcraft. Latour and Woolgar come close to confessing outright this sort of debunking view of science as witchcraft in a footnote: "One particularly useful source [for their own study] was Auge 's analysis of witchcraft in the Ivory Coast, which provides an intellectual framework for resistance to being impressed by scientific endeavour." Just as the field anthropologist never really believes that the tribal sorcerer under study is actually breaking the laws of nature with his or her witchcraft, so, according to Latour and Woolgar, the sociologist of science ought never to believe the scientists' own claims that science finds out "the facts," the "actual way things really are." As Latour and Woolgar see it, the task of the sociologist of science is to give a debunking account of scientific practice, one that shows how practitioners of science become deluded into thinking that they are discovering a reality that was already in existence rather than constructing a reality by artificial means (which is actually what they are doing instead). It is no accident therefore when Latour and Woolgar admit unashamedly that they studied the behavior of the scientists in the lab as though those scientists were the postindustrial equivalent of, and here I want to use the very expression Latour and Woolgar use, *primitive sorcerers* engaged in routinized behavior that was entirely "strange" and virtually incomprehensible to Latour and Woolgar. One bizarre oddity to their study (which the intelligent reader can hardly miss) is how "strange" Latour and Woolgar seem in failing to understand or appreciate the obvious explanation for this strangeness they felt toward what they were observing in the lab: Namely, unfamiliarity and lack of experience with *any* kind of activity will tend to make that activity seem strange when seen from the outside. But instead of condemning their own ignorance as scientifically naive observers, Latour and Woolgar conclude that they are the only people in the lab capable of an objective understanding of what the scientists are doing. Ignorance is a necessary condition of wisdom, apparently, in the ironic world of the sociology of science.

What those scientists are doing is not what those scientists think that they are doing, according to Latour and Woolgar. The field study is rather humorous to read, although some of the humor is surely unintentional on Latour and Woolgar 's part. The two field anthropologists are thoroughly perplexed by the daily activities of the lab workers, activities that the sociologists describe in the most observationally literal way they can invent. Machines hum and churn and produce "wavy lines," which are then superimposed; flasks of liquid are shaken, then argued over; workers consult texts in order to produce more "inscriptions." The attempt is to describe moment-to-moment practice in a bioscience laboratory in such a manner as to make it out to be eccentric and arbitrary enough to qualify as irrational: science as a more fancy form of witchcraft. What truly shocks Latour and Woolgar is the claim made by the practitioners they are studying that they, the practitioners, are doing something much more worthwhile than arbitrary witchcraft that they are discovering objective structures in the domain of nature under their investigation, that they are finding out objective facts about neurohormones. Latour and Woolgar are sure that this is not so, that any claim to have discovered preexisting material realities is a self-congratulatory form of exaggeration on the part of the scientists, basically nothing more than arrogant hubris; for all that is *really* going on in the lab is that, machines hum and churn and produce wavy lines, flasks of colored liquid are shaken, then argued about, workers consult texts in order to produce more inscriptions, and so on.

The problem seems to be that Latour and Woolgar make a basic mistake in the epistemology of science. They take the fact that, often in science, the detection of an objectively existing structural reality requires an elaborate detection procedure (involving elaborate machines and so on) to mean that no such objectively existing structural realities exist; for what the practitioners are "detecting" in such cases is an artificial structural reality, one created by the elaborate detection procedure. In other words, for Latour and Woolgar, The wavy lines produced by the fancy machine that produces wavy lines are not wavy lines *about anything except the very process that produced them*. The content of science is merely the process by which it is artificially constructed as knowledge it is not the structure of some independent reality. No properties of an independent reality are represented in the wavy lines, for the wavy lines are entirely artificially manufactured by the machine that makes wavy lines. When the scientists themselves object to this that the wavy lines contain peaks and valleys which represent some of the chemical properties of a hormone sample that was put into the machine that makes wavy lines, Latour and Woolgar will believe none of it; for that is too much like listening to the tribal witch doctors insisting that the spell they cast really did cure the sick daughter of the assistant chief by driving away the independently existing evil spirits.

Latour and Woolgar will have none of it in this way because they make a specific philosophical mistake: From the fact that our *being able to say or to claim* that X exists depends on an elaborate and artificial procedure for detecting X, they mistakenly think it follows that X's existence itself depends on that elaborate and artificial detection procedure. But, of course, this is an absurd as well as an invalid inference. To see the absurdity, consider the following implication of the Latour and Woolgar view. We cannot observe viruses without very elaborate and wholly artificial equipment such as electron microscopes, special chemical stains, and so on. If Latour and Woolgar are correct, that would imply that the existence of viruses *depends* on such artificial equipment and such elaborate detection procedures. If such procedures and equipment did not exist, *then viruses would not exist.*

Why do Latour and Woolgar conceive the sociology of science to be this kind of debunking project? What motivates their confidence that practitioners of modern science are as serenely mistaken about the real nature of their own business as are the tribal witch doctors about the real nature of their business? Well, Latour and Woolgar claim that the facts indeed, all the facts of science are entirely socially constructed. Scientific facts are really about their own social construction, not about some independent reality. But they realize that everybody thinks scientific facts are about some inquiry-independent reality. Now most scientists are fairly smart. What would lead smart people to make this sort of mistake about the real status of scientific facts? One central chore of the sociology of science must be to explain how it is that what is entirely socially constructed and about one sort of subject matter is almost universally mistaken by otherwise bright human beings to be not entirely socially constructed and about a completely different kind of subject matter. This will force the sociology of science to become primarily a project of debunking. Latour and Woolgar attempt to provide the missing explanation of the mistake: The methods of science are designed to "cover up" the social construction of the products of science. Once a purported fact has been established by scientific means, the elaborate and artificial procedures used to establish it conveniently disappear from notice and are forgotten. The fact now seems to "stand on its own"; hence, the delusion (Latour and Woolgar 's word for it is 'illusion', but they are being sloppy here-they mean to speak of what is more properly called a delusion) naturally arises that the fact is about an independently existing objective reality and not about its own process of construction through organized social behavior. That, according to Latour and Woolgar, is why the wavy lines produced by the machine that makes wavy lines are universally held to have been made by the interaction of an independent reality with the machine, when actually they were complete creations of the machine and how it was manipulated by its users.

How does the "cover up" occur? Here Latour and Woolgar become very unclear; for, despite their lack of empathy with philosophy, they use a central philosophical notion in their account of the covering up process: the notion of a modality. In modal logic and metaphysics philosophers and logicians have worked out relatively precise characterizations of various modalities. Unfortunately, Latour and Woolgar are completely innocent of this prior work. They don't quite mean by 'modality' what philosophers and modal logicians mean by 'modality'. Just what they do mean is not entirely clear. A modality is a "mode" that qualifies the truth-conditions of a sentence. For example, 'possibly' is taken to set up a modal context. So, the sentence,

UFOs are spacecraft piloted by alien beings.

does not mean the same (because it is not true or false under *exactly* the same conditions) as the sentence,

Possibly, UFOs are spacecraft piloted by alien beings.

Even if we knew that the first sentence is false, the second one could still be true that's the difference that 'possibly' makes to the truth-conditions for what would be otherwise the same sentence. There are many kinds of non modal sentences that nonetheless function logically like modalities that the philosopher Bertrand Russell called propositional attitudes. Propositional attitudes have to do with the different psychological attitudes that a human being can take "toward" a sentence (actually, the attitude is taken toward the content of the sentence, which philosophers call the proposition of the sentence). For example, the sentence,

Some UFOs are spacecraft piloted by alien beings

does not mean the same as, indeed, it has very different truth-conditions, than the sentence,

Harvey is skeptical that some UFOs are spacecraft piloted by alien beings.

And both those sentences are yet different in meaning from the sentence,

Harvey is confident that some UFOs are spacecraft piloted by alien beings; which is different again from the sentence,

Harvey wishes it weren't commonly believed that some UFOs are spacecraft piloted by alien beings.

I have put all the different propositional attitude indicator phrases in italics in the above examples. The reader can see the basic pattern. The same declarative sentence occurs after a different attitude phrase ending with the word 'that'. This is the commonly accepted way of representing propositional attitudes. Notice that the truth-value of the buried declarative sentence can stay the same while each sentence as a whole can change truth-value depending on the content of the attitude phrase. Suppose, for example, that as a matter of fact it is false that some UFOs are spacecraft piloted by alien beings. Nevertheless, it can be true that Harvey is skeptical of this fact, false that he is confident of that fact, and true that he wishes this falsehood weren't so commonly believed to be true by so many other people. The reader can now sense the virtually limitless number of different propositional attitudes: 'believes that', 'fears that', knows that', 'doubts that ...', and so on, are all propositional attitudes in which a declarative sentence fills the position. Here is the important point about propositional attitudes for purposes of our present discussion: The truth or falsehood of a propositional attitude sentence is not purely a function of the truth or falsehood of the embedded declarative sentence rather, the specific content of the attitude factors in also.

Latour and Woolgar are misled by the presence of propositional attitudes in scientific reasoning and practice into the belief that all scientific facts are about their own construction, not about some practice-independent reality. What misleads them is how different propositional attitudes occur at different times in the process of discovering a scientific fact. The attitudes expressed early in the discovery process tend to be attitudes involving skepticism and uncertainty. Surely this ought to be no surprise. Equally, it ought to be no surprise, that if scientists are fortunate and their conjectures are true, nature rewards their researches with success, and they begin to express attitudes toward the same "fact" that involve increasing confidence and sureness. For example, a research paper appearing in an immunology journal in 1987 might contain the following propositional attitude sentence, *We now suspect that* NK cells kill some of their targets with the aid of antibodies.

This sentence indicates a tentative attitude toward the alleged fact in question. It is being surmised, floated as a possibility that subsequent practice will either support or reject. We can imagine it happening that by 1998 the same embedded sentence appears in a research paper attached to a more confident attitude, *We now believe beyond doubt that* NK cells kill some of their targets with the aid of antibodies.

Latour and Woolgar take this evolution of attitudes from uncertainty to confidence to show that the belief scientists have that they are discovering preexisting material realities is a delusion; for, according to Latour and Woolgar, the move to more confident attitudes is a purely social creation, one having to do with how many times a given paper is cited in other papers, in textbooks, and in conversation, how effectively the prestige of one scientist can intimidate others into accepting his or her views, how standardized lab machinery becomes so that the same "wavy lines" are reproducible in different labs, and so on, and not one having to do with uncovering a fixed material reality. Their argument is instructively erroneous enough to be worth quoting at length:

> *In simple terms, scientists were more convinced that an inscription unambiguously related to a substance "out there," if a similar inscription could also be found. In the same way, an important factor in the acceptance of a statement was the recognition by others of another statement that was similar. The combination of two or more apparently similar statements concretised the existence of some external object or objective condition of which statements were taken to be indicators. Sources of "subjectivity" thus disappeared in the face of more than one statement, and the initial statement could be taken at face value without qualification. An "object" was thus achieved through the superimposition of several statements or documents in such a way that all the statements were seen to relate to something outside of, or beyond, the reader's or author's subjectivity.*

External objects are said to be "concretised" inventions due to the repeatability of a given "statement" in the research literature. Objects are said to be "achieved" as constructions out of socially produced data that are repeatedly obtainable only because of the standardization of research equipment and procedures. It is no wonder then that later in their study Latour and Woolgar call the objectivity of the concretised, achieved material reality a mere "impression of objectivity". It isn't real objectivity at all, is the clear implication. The scientists are as deluded as the tribal witch doctors. If this isn't antirealism by the case method with a vengeance, one wonders what would be.

It is important for us to understand what has happened to the allegedly objective world in the hands of Latour and Woolgar. Here again, we see at work the pernicious influence of the Kantian distinction between phenomena and noumena. *The noumenal world is taken to be so epistemologically remote by Latour and Woolgar as to be never represented in the output of our detecting devices at any stage of our detecting procedures.* Put a slightly different way that makes clearer the antirealism, *the amount of "filtration" through artificial manipulation that scientific data undergo during collection is so great as to destroy whatever was represented about the objective world in the original input at the distal end of the data collection process.* We might call this "The Great Filtration" or "The Great Distortion." It has a distinctly Kantian flavor to it. If one seriously believes in The Great Filtration/Distortion, then it will follow as a matter of course that science is about its own process of construction rather than about an objective world. Again, I quote Latour and Woolgar at length:

> *Despite the fact that our scientists held the belief that the inscriptions could be representations or indicators of some entity with an independent existence "out there," we have argued that such entities were constituted solely through the use of these inscriptions. It is not simply that differences between curves [on superimposed graphs] indicate the presence of a substance; rather the substance is identical with perceived differences between curves. In order to stress this point, we have eschewed the use of expressions such as "the substance was discovered by using a bioassay" or "the object was found as a result of identifying differences between two peaks." To employ such expressions would be to convey the misleading impression that the presence of certain objects was a pregiven and that such objects merely awaited the timely revelation of their existence by scientists. By contrast, we do not conceive of scientists using various strategies as pulling back the curtain on pre given, but hithereto concealed, truths. Rather, objects (in this case, substances) are constituted through the artful creativity of scientists.*

One hardly knows what to make of the claim that a substance is "identical" with perceived differences between superimposed graphed curves. Bloor promised that the sociology of science would not be a form of idealism, but Latour and Woolgar in the above quote would seem to be wavering on the precipice of idealism. Substances are said to be identical with certain perceptual experiences. One doesn't know whether to laugh or cry when Latour and Woolgar tell the story of TRF (thyrotropin releasing factor, a neurohormone) as though TRF, the actual physical hormone itself (not merely facts about it), was created; "a decisive metamorphosis occurred in the nature of the constructed object" *Laboratory Life*) sometime between 1962 and 1969 in two particular laboratories.

Latour and Woolgar claim that the upshot of their particular brand of social constructivism is that *reality is the consequence of the content of science, not its cause.* Because they hold this view, they object to the idea that one could explain why different scientists arrive at similar or convergent beliefs by appealing to an independent world whose stable structure is the same for all scientists; rather, Latour and Woolgar insist, a common independent world is the entirely constructed outcome of the practices of scientists. It follows that realism gets it backward: The structure of the world doesn't constrain the content of science, the content of science constrains the structure of the world. The realist philosopher of science Philip Kitcher argues that for Latour and Woolgar to make good on this last claim they would have to show that differences in scientists' "encounters with the world" (read, with a noumenal, independent material reality) would make no difference to the outcome of what counts as the scientific facts, that no matter how those encounters with the world had been different, scientific facts would have been constructed with the same content. This, Kitcher argues, the two sociologists of knowledge fail to show with their case studies. I might add that what those case studies do show is that differences in the constructed content of scientific facts make a difference to the quality of our encounters with the world. But one could find enlightened realists who would agree with that. If Kitcher is correct, then Latour and Woolgar 's interpretation of their study done at the Salk Institute is underargued and a rather misleading exaggeration.

In Latour and Woolgar's view, far from science being constrained in its content by a preexisting world "out there" whose structure it slowly discovers, science is a self-justifying collection of arcane technological practices replete with mystifying jargon, guild-like exclusivity of participation, and all the trappings of ideological self-center ednessa collection of arcane technological practices that literally invents the world from scratch. This is nuclear-strength antirealism if anything is.

Latour and Woolgar fess up to their antirealism eventually when they get around to a brief discussion of who they regard as the opposition. They identify the enemy of the sociology of knowledge to be what they call "a realist theory of science". Their argument against realism is so woolly as to be nearly unrecoverable, but their basic idea is that realism is a delusion. Their main reason for holding that realism is a delusion seems to be the old howler of a view that every philosophy teacher since the formal teaching of philosophy began has encountered coming out of the mouths of novice students in introductory-level philosophy courses: Namely, because you can't talk about an objective and independent material world without using concepts formed during the socially determined investigation of such a world, there is no such world at all there are only the socially determined practices of investigation that produce the said concepts and the talk we make with them. The mistaken reasoning here is blatant and irreparable.

Latour and Woolgar 's argument: If there were an independent material world, it would have to be directly knowable through nonlinguistic, non conceptual means. But the second we start to talk about what we claim to know of such a world, we are helping ourselves to socially constructed concepts; and, if the concepts are socially constructed, so also are what they are concepts about.

The last clause in this argument is the sort of mistake that ought to remind the reader of one of Bishop Berkeley's more embarrassing moments: Because you can't think that trees exist independently of being thought about without, like, *thinking* about them, there are no such trees there are only thoughts about trees; or rather, trees "just are" nothing but thoughts about trees. The way out from under this woolly argument of Latour and Woolgar 's is surely pretty obvious: The first sentence of the argument is just plain false. Who says that if there is an independent material world it must be knowable directly by nonlinguistic, non conceptual means? Certainly no realist would make such an implausible demand. Only the Kantian idealist would think up such a self-defeating and question-begging requirement. But Latour and Woolgar, if they are anything philosophical at all, are true to their Kantian (hence, to their idealist) roots: For them what exists collapses into what is knowablein good Kantian idealist fashion-precisely because for them what exists collapses into what is investigatable, and what is investigatable collapses into what is describable, and what is describable collapses into what is knowable. In this tortuous way, Latour and Woolgar lead the sociology of knowledge up a dead-end alley into full-blooded antirealism, an antirealism that comes perilously close to what Bloor said the sociology of knowledge must avoid: an idealist model of science.

View from Deep in the Swamp of History

It was inevitable that Kuhn's historicizing of philosophy of science would lead to a host of extremely detailed studies of particular episodes in the history of science. Book-length studies were surely destined to appear in which scholars would meticulously pour over the small-scale historical trivia of science: Galileo's private letters to his mistresses, Einstein's eating habits between thought-experiments, Darwin's sexual inhibitions as revealed by remarks in his journal on the breeding of domestic animals. I do not know whether any treatises have been published on the three above topics, but one social constructivist historical study did appear that caused a bit of a stir: Steven Shapin and Simon Schaffer's examination of the dispute between Robert Boyle and Thomas Hobbes over Boyle's air-pump and what it allegedly showed about the role of experimentation in science, a dispute that started in 1660 and lasted until Hobbes's death in 1679. The standard view of this dispute makes Boyle out to be the hero and Hobbes the villain. Boyle's position is taken to be the correct one while Hobbes's position is often portrayed as that of a boorish crank's, as though Hobbes was a mere armchair philosopher (whereas Boyle was a laboratory scientist) pathetically railing against the legitimacy of scientific experimentation from his moldy apriorist perch.

Shapin and Schaffer beg to differ. They seek in their study to show that Boyle didn't know as much as he has been thought to have known, that Hobbes was not quite the boorish crank the standard view says he was, and that Boyle's victory in the dispute was not due to any genuine superiority of his position vis a-vis the objective facts about the physical world. Hobbes's rejection of Boyle's interpretations of certain experiments with his air-pump was cogent on its own terms, claim Shapin and Schaffer, and the wrongness of Hobbes's position is not much more than partisan history of science as written by the followers of Boyle. That is certainly an ambitious goal to aim their study at, and if Shapin and Schaffer could pull it off, then we should all be grateful for their having rectified a huge mistake in the heretofore accepted understanding of a critically important episode in the history of science.

In the mid seventeenth century science was not yet understood as an empirical form of inquiry clearly distinguished from traditional philosophy. In fact it was common to refer to what we would now call science by the label "natural philosophy." As an example of how mingled together science and philosophy were, the philosopher Thomas Hobbes in his main work published in 1655 defined philosophy as the study of the causes of things. No philosopher nowadays would find that definition plausible at all: Philosophy is not a causal enterprise, it is an analytical form of inquiry, one that seeks to clarify, systematize, and criticize the concepts and methods of other more specialized domains of inquiry. The distinction we now take for granted between issues involving the conventional definitions of words and issues involving matters of fact about physical reality was not nearly as clear in 1660, even to the best minds of the times. It was a time when that distinction was first being worked out by human beings.

Robert Boyle was a brilliant natural philosopher who championed a new way of gathering knowledge, a way that did not fit the traditional form knowledge gathering took in established philosophy. Boyle was suspicious of grand a priori systems, of philosophers who presented complicated deductive systems of definitional truths that the structure of the world must supposedly match because reason itself supposedly necessitated such a match. A grand a priori system was precisely what Hobbes had labored to produce over many years and presented to the world in his monumental *De Corpore* in 1655. Against such overbearing systems of metaphysics Boyle advocated something brand new: the experimental matter of fact. Where the former kinds of metaphysical systems were all doubtful because they were entirely speculative, the latter facts of the matter were certain because they were directly witnessed through ordinary sense perception. Boyle made a distinction between knowing what a matter of fact is and knowing what its cause is. The former kind of knowledge is available to inquiring humans who perform the right sorts of experiments, the latter sort of knowledge is largely forbidden humans because it is beyond their god given means of perception to ascertain the hidden structures of things. Hobbes was of course aghast at such a distinction, for it meant that properly philosophical knowledge of causes was for the most part inaccessible to human beings. Hobbes was suspicious of experimentation, of mechanically inclined thinkers who invented an artificial machine and then claimed to have discovered this and that about the world on the basis of the machine's behavior under artificial conditions.

Obviously, bad blood was bound to flow between Hobbes and Boyle. Boyle invented an air-pump, a machine he claimed evacuated almost all the air from an enclosed receptacle. He published a book reporting the results of forty-three experiments performed with the air-pump. This was a watershed moment in the history of science; for with Boyle's air-pump we see the arrival on the scene of two permanent institutional features of modern science: the standardized physical laboratory as a public space for experimentation, and the official skepticism toward all things theoretical (as opposed to all things observational) that became a universal requirement of the scientific mind-set so universal and so required that Karl Popper later made it the very heart of scientific method with his falsificationism. Boyle's air-pump, as Shapin and Schaffer show, was the "big money" science of its time, the seventeenth-century equivalent of a modern day subatomic particle accelerator or a space shuttle. So fancy was it, so difficult and expensive to design and produce was it, that no more than four or five working air-pumps ever existed in the entire world during the first decade of the Boyle/ Hobbes dispute. By modern standards, of course, making Boyle's air-pump would be a very achievable project for a bright student competing in a school science fair. No matter, Boyle's air-pump was once the center of an intellectual firestorm during the birth pangs of modern science. Boyle claimed that some of his experiments with the air-pump proved certain incontestable "matters of fact": that the air has a "spring" to it, that the air has a weight to it different from its spring, that a vacuum is possible in nature, and so plenism (the philosophical position that a vacuum in nature is impossible on a priori grounds) is false. Hobbes was an "anti-vacuist," one of the armchair philosophers who held that a vacuum was impossible on grounds of pure reason. Naturally, Hobbes objected to Boyle's claim that the air-pump experiments involved the creation of a vacuum or near-vacuum inside the pumped receptacle. Shapin and Schaffer present the subsequent dispute in all its floridly bitter detail. A host of side characters appear in the lengthy drama, among them the Dutch natural philosopher Christian Huygens, Robert Hooke, a Jesuit priest named Franciscus Linus, and Henry More. Round and round they bickered and argued, experimented and counter experimented (accept for Hobbes and his fellow philosopher Spinoza who both scoffed at experimentation as inferior to pure reason), and wrote treatises, tracts, and diatribes. Hobbes insisted that Boyle's air-pump leaked indeed, that it *must* leak on a priori grounds for a vacuum requires self-moving matter, which is a self contradiction. Boyle insisted that the air-pump's leakage was negligible, that it created a vacuum for all *practical* purposes, and that the vacuum it created was a certain matter of fact on which all experimentalists could agree because it was publicly witnessable and repeatable. Nothing less than what form of life science would take was at stake, according to Shapin and Schaffer: Would it be the experimental form of life, or the form of life of a priori metaphysical system building? We can know only observational facts for sure, said Boyle, we are restricted to mere speculation about the underlying causes. Knowledge is of causes and how they fit into a grand metaphysical system or it is of nothing, said Hobbes, the so-called observational facts are mere artifacts of the artificial machines used to create them.

Inequality of Theoretical Alternatives

Shapin and Schaffer's study is interesting and important. It shows just how conceptually messy science can be, especially when it was in its infancy and nothing could be taken for granted about it. The two historians take the under determination of theory by observational data as an established given throughout their analysis of the Boyle/ Hobbes dispute. What is puzzling in their analysis is the lack of any recognition on their part that just because there exist alternative theoretical explanations for a given collection of observational data doesn't mean that all those alternative explanations are equally plausible and equally likely to be correct. Shapin and Schaffer seem to believe that there can be no principled way of ranking competing alternative accounts of observational data; therefore, they take it as established that *if* there are, say, six alternative and mutually incompatible theoretical explanations of a particular observational datum, there can be no question of one of those six theoretical explanations being any better than the other five. But why should anyone believe that? *Not all theoretical alternatives are equal.* Some alternative explanations cost too much in terms of changes we would have to make in other beliefs. Some alternative explanations postulate new objects or properties that we have no other reason to suppose exist, and so the postulation merely to explain this one observational data-set is *ad hoc*. Some alternative explanations increase the structural complexity of the system under investigation over what other alternative explanations do, and so the former explanations run a higher prior probability of being false. The ways in which competing alternative theories can be unequal with respect to their consequences for methodology, epistemology, and ontology are legion. Shapin and Schaffer seem unaware of this feature of competition between theories. They jump from the fact that Hobbes offered an alternative theoretical account of Boyle's air-pump experiments to the conclusion that Hobbes's account is just as plausible as Boyle's account. But there is no reason to suppose this kind of equality exists among competing theories. In fact, we know that it is extremely difficult to show that two theories are *exactly observationally equivalent*. The only known cases come from rare episodes in the history of physics. But even if two theories can be shown to be exactly observationally equivalent, it doesn't follow from such equivalence alone that the two theories are equivalent with respect to their methodological, epistemological, and ontological implications. We saw in chapter 4 how Quine, who is very fond of the under determination of theory by observational data, nevertheless presents a host of principles that can be used to rank competing theories: simplicity, conservatism, modesty, and fertility. Arguments can be given for why theories that maximize simplicity, consistency with our established beliefs (conservatism), modesty, and fertility at generating novel predictions are more likely to be self-correcting toward the truth over the long run than theories that do not maximize those pragmatic virtues. Hobbes's alternative account of the air-pump experiments did not win out over its rivals in the end. In fact, no theories that postulated a subtle fluid aetherof which Hobbes's was a variety survived into the twentieth century. They had too many other epistemological costs that would have made it unwise to keep them afloat. Shapin and Schaffer conveniently leave out this subsequent history of atmospheric science. The suggestion at the heart of their study is that if there had been a few key differences in the sociopolitical factors that made Boyle's view triumph, then we would be Hobbesians this very day as you read this that Hobbes's theories failed not because they were false but because the contingent politics of early science went against them.

Scientific truth is a function of human sociopolitical intrigue much more than it is a function of what the objective structure of the world is like, is Shapin and Schaffer's point. As they say near the end of their study, it is "not reality which is responsible for what we know". Once again, the principle of the under determination of theory by observational data is being asked to support a form of antirealism that it does not logically entail. This argumentative maneuver is a standard strategy of social constructivists in every discipline. Their assumption seems to be that if nature itself does not unambiguously point to a unique theoretical explanation for something, then there can be no principled way of claiming that there can be an objective truth about that something. But this conditional is unfounded. We can now say with supremely high confidence that:

1. Boyle was correct that the air has weight.
2. Boyle was correct that vacuums are possible.
3. Boyle was incorrect that the air has "spring."
4. Boyle was incorrect that the air contains a subtle aether.
5. Hobbes was incorrect in certain claims about the air (it contains a subtle aether, vacuums are impossible, and all chemical phenomena are mechanical in nature).

Shapin and Schaffer would apparently have us believe that these truths are neither objective nor apolitical; and Latour and Woolgar would add that the above truths only *seem* to be objective and independent of sociopolitical determination because the official mythology of science covers up their socially constructed pedigree. It is neither true nor false, per se, that the air has a spring, according to the social constructivists at best, it is false *relative* to the sociopolitical framework of our current scientific form of life. But, had that sociopolitical framework and its attendant form of life been different in certain ways, then it would be true that the air has a spring. This last claim might strike others besides myself as more than just a little fantastic. But if it were true, then the science critic is potentially greatly empowered to change society through changing its dominant form of knowledge gathering: science. If the facts *all* the facts-can be made different by making the social conditions of their construction different, then it is a short step for those social constructivists who harbor an activist streak to see their critiques of science as a form of revolutionary social activism. That is one reason why there is a certain kind of science critic who doesn't mind a model of science in which the input of an independent material world has been reduced to insignificance or eliminated altogether, who is in fact very glad to have arrived at that result. For, to such a science critic, the end product of scientific endeavor under such a model of science is entirely or almost entirely a function of complicated social behavior; and, given that the content (or almost all the content) of science is therefore determined socially, that content is potentially alterable by altering the social conditions of its production. Thus, the possibility arises of altering the substantive content of science through political activism that seeks to alter the social conditions under which the production of scientific knowledge takes place. Some social constructivists, such as Bloor, are not specifically interested in this further activist aspect of the sociology of knowledge. They seem satisfied to point out the social determination processes by which the content of science is produced. We might describe this by saying that those social constructivists stress the *social* in social constructivism. Other social constructivists, in particular the feminist science critics of the next chapter, stress the *constructivism* aspect of social constructivism, the view that the factual content of science can to a considerable extent be constrained by self-conscious social engineering, by deliberate intervention in the scientific process for purposes of furthering a partisan political agenda. What I want to suggest here is how the non feminist social constructivist accounts of science naturally lead, via openly idealist variants of the sociology of knowledge, to the more radically interventionist feminist models of science.

16

Foundations of Mathematics and its Founders

Introduction

Many of the greatest mathematicians and philosophers have set themselves the task of finding a 'proper foundation,' either for mathematics or for some special branch of mathematics. They have declared the need to give clear definitions of mathematical concepts, to formulate rigorous proofs and to trace mathematical theorems to genuine first principles. But this consensus hides not only considerable divergence about how clear definitions, full rigour, and genuine foundations are to be achieved, but also differences in motivation. There are several quite distinct reasons why the enterprise of exposing the 'proper foundations' of mathematics has seemed important to mathematicians and philosophers, and we cannot hope to understand the activity or its history without appreciating these reasons.

The most obvious incentives for attending to definitions, proofs and first principles come from identifiable difficulties in the practice of some branches of mathematics. If mathematicians are employing problem-solving techniques that are recognisably powerful in some cases but that yield palpable errors or, worse, paradoxes in others, they will want to clarify the concepts they are using or scrutinise their methods of reasoning. History shows that mathematicians have frequently investigated the foundations of their subject precisely because they find it impossible to carry forward their technical research without clarifying concepts or modes of reasoning.

However, foundational activity also arises from other motivations. Sometimes mathematicians hope to enhance their understanding of the results in a particular mathematical field by making the interconnections as explicit as possible; they try to systematise the field and to make clear exactly what depends on what. If successful, the attempt may bring improved understanding even if there was no prior paradox or error. In setting a discipline in order, mathematicians are effectively behaving like natural scientists who seek explanations by trying to develop a unifying theory that will systematise a disconnected collection of empirical findings.

Finally, there may be purely philosophical motives for attending to the foundations of mathematics. For many thinkers, mathematicians and scientists as well as philosophers, mathematics has served as the paradigm for human knowledge because of the apparent certainty of its conclusions. The prior conviction that mathematical knowledge is absolutely unshakeable may inspire efforts to show exactly how this certainty is obtained. Alternatively, there may be concern about some particular area of mathematics which is seen as falling short of the indubitability that mathematical knowledge ought to possess, such that researchers try to remedy the defect by pursuing the subject's foundations. In either instance, the goal will be identical first principles will be identified that are immediately certain and independent of sensory experience, and chains of reasoning will be developed that depend only on these principles as premises and proceed by inferential steps that preserve certainty.

It is important to distinguish these three main forms that the search for the foundations of mathematics may take if we are to see why some enterprises have succeeded and others failed. Mathematicians can achieve *problem-solving foundations* (foundations of my first type), which enable them to carry forward their technical research, without having either *explanation-yielding foundations* (my second type) or *certainty-conferring* foundations (my third type). Similarly, an endeavour to uncover explanation-yielding foundations may succeed even though the first principles that it identifies are not known immediately and for certain. Unfortunately, through rhetoric and lack of philosophical precision many earlier writers blurred the above distinctions. Moreover, while mathematicians sometimes write prefaces as if they were concerned to establish foundations of all three types, their practice usually belies the advertisement.

Early Searches for Foundations

Although mathematics has ancient roots in the civilisations of Mesopotamia, the first deductive organisation of the subject known to us is the *Elements* of Euclid (*c.* 300 BC). Euclid's presentation starts with some definitions, postulates and 'common notions', and on this basis, Euclid derives the main results of the geometry of plane linear figures and circles. Euclid does not, however, inform the reader what this presentation of the *Elements* was intended to achieve. Commentators on his work are more forthcoming, but they use ambiguous language and fail to distinguish between the various kinds of enterprise outlined above. Hence Euclid's work was able to serve as the paradigm for different types of foundational programme: he could be read as tracing the theorems of geometry to immediately certain first principles about the structure of space, or as showing how it is possible to provide an explanatory organisation of known mathematical results, without any commitment to the certainty of the first principles he adduced.

Despite Euclid's profound influence both on Greek geometers and on the mathematicians of the Renaissance and after, there was a flourishing mathematical tradition that did not receive the foundational treatment exemplified by the *Elements*. Babylonian mathematicians had discovered methods for solving practical problems that we would formulate using linear and quadratic equations, and the sciences of arithmetic and elementary algebra were carried forward during the medieval period by Arab and Indian mathematicians. The tradition continued to thrive in renaissance Italy, where the practice of challenging the skills of others stimulated several mathematicians to investigate questions that are, in our terms, tantamount to solving cubic equations. In 1545, Gerolamo Cardano (1501–76) published his *Ars magna*, in which he describes methods for solving various types of cubic equation, and also briefly mentions a foundational question.

Even before the explicit formulation of algebraic notation, workers in the theory of equations recognised that some problems afforded a smaller number of solutions than was normal. The defect could be remedied by allowing strange entities to count as solutions, such as 'negative numbers' or 'numbers whose square is less than zero', but there was considerable suspicion of the propriety of doing so. The Italian algebraists saw (in effect) that they could obtain exactly n roots for an equation of nth degree, but this seemed a slight benefit compared with the cost of dabbling in mysteries. In Cardano's phrase, such 'numbers' are 'subtile and useless'.

However, the seventeenth and eighteenth centuries were to witness a transformation of algebra and the flowering of algebraic analysis. Pierre de Fermat (1601–65) and Rene Descartes (1596–1650) independently showed how the algebraic notation of François Viète (1540–1603) could be applied to the solution of geometrical problems. Their new analytic geometry was able to generalise and extend the results of ancient Greek geometers (paradigmatically, Apollonius on conic sections). Building on the work of Descartes, Fermat and other mathematicians, Isaac Newton (1642–1727) and Gottfried Leibniz (1646–1716) independently formulated the principles of the differential and integral calculus. Using the new calculus, Newton and Leibniz computed the values of the areas under many curves, found the lengths of arcs, identified radii of curvature, constructed tangents and normals and solved problems about the motions of ideal particles. The Marquis Guillaume de l'Hôpital (1661–1701), author of the first textbook on the Leibnizian calculus, claimed that the new methods enabled mathematicians to solve problems which 'nobody previously had dared to attempt'.

Unfortunately, the new successes were purchased at a cost. In differentiation and integration, the fundamental operations of the new calculus, both Leibniz and Newton were compelled to use language that appeared to refer to illunderstood infinitesimal entities. The procedure for differentiation makes the trouble clear. If $y = f(x)$, we compute the quantity dy/dx by forming the difference $f(x + o) - f(x)$ and dividing by o. We write the result as $A + B[o]$, where the latter term contains all and only those expressions that are powers in o. Finally, we neglect terms in o by simply dropping $B[o]$ and declaring that $\mathrm{d}y/\mathrm{d}x = A$.

It is easy to appreciate why this (and kindred manipulations) should cause qualms. For if o is not zero, then by what right do we omit those terms that are powers in o? On the other hand, if $o = 0$ then the initial difference $f(x+o)-f(x)$ is also o, and, perhaps more alarming, the procedure involves a division by zero. Hence there seems to be no way of justifying the algebra or of understanding the operations that are supposed to be performed. Early critics of the calculus, most notably George Berkeley (1685–1753), made the objection with considerable force. Nonetheless, the new methods were enormously fruitful, for, in the hands of Jakob Bernoulli (1654–1705), John Bernoulli (1667–1748) and Leonhard Euler (1707–83), they yielded conclusions that were both recognisably correct and significant in finding answers to traditional mathematical questions.

The successes of the calculus were not initially hampered by these difficulties and mathematicians had no motive to seek problem-solving foundations for the new algebraic analysis. However, in the works of Newton, Leibniz and their successors, there are remarks that can be interpreted as indicating either a desire for explanation-yielding foundations or for certainty-conferring foundations. Newton and Leibniz both pondered the question of whether the calculus could be reconstructed after the manner of 'the Ancient Geometers' (most prominently Euclid). Newton's preferred method of formulating the central concepts of the calculus embodied a kinematic conception of geometry (on which curves are viewed as generated by the motion of a point), and he articulated a justification of his conclusions about differentiation that we can identify, with hindsight, as a cryptic version of an appeal to the notion of a limit. Leibniz's published writings on the calculus appear pragmatic by contrast, for he seems willing to allow that the algebraic work is autonomous and that it does not need an interpretation in geometric or kinematic terms. The contrast is heightened in the works of eighteenth-century Leibnizians and Newtonians, for while Newton's successors continued to devote considerable energy to showing how the Newtonian calculus could be rendered according to the ideas of traditional geometry, the followers of Leibniz were more concerned to extend the scope of the new discipline.

During the eighteenth century, the issues surrounding the foundations of algebraic analysis became more complicated. In their attempts to differentiate and integrate recalcitrant functions, both

Newton and Leibniz had introduced infinite series expansions and had used term-by-term integration and differentiation.

Leibniz's method sometimes generated peculiar results, showing that the techniques had to be used cautiously. Nonetheless, in the hands of Euler, algebraic manipulations of infinite series expressions led to surprising discoveries. Euler recognised the distinction between convergent and divergent series, and he handled the latter with enormous artistry. Moreover, following the lead of Leibniz, Euler used complex numbers in the evaluation of indefinite integrals and he extended the work of Roger Cotes and Abraham de Moivre by giving a systematic treatment of the relations between exponential and trigonometric functions. Complex numbers could no longer be dismissed as 'subtile and useless' for they had a firm place within algebraic analysis.

Yet scholars with foundational worries could still protest that, for all its successes, algebraic analysis failed to live up to the standards of genuine mathematics. Infinitesimalist reasoning seemed only to work by luck, infinite series manipulations had to be performed with care (or with 'intuition'), and the reference to negative and complex numbers remained mysterious. Periodically during the eighteenth century mathematicians returned briefly to the foundational issues; but in retrospect, we can see these as mere uncoordinated groupings towards the major ideas of the nineteenth century. In my judgement, the exciting achievements of algebraic analysis made a search for problem-solving foundations idle, and, while explanation-yielding foundations would have been a bonus, the prize was too small to divert creative mathematicians from extending the scope of analysis. So the recurrent lament that mathematics no longer accorded with the standards of Euclid did not translate into persistent foundational activity.

Nineteenth Century

Geometers, however, were attending to the substance of Euclid's *Elements*. Many mathematicians had found that the fifth postulate was less evident than the other four, and, in the centuries since Euclid, there had been periodic attempts to prove the fifth postulate, attempts which typically made use of equivalent principles. Following earlier, less careful discussions, Carl Friedrich Gauss

(1777–1855), Janos Bolyai (1802–60) and Nicolai Lobachevsky (1792–1856) all recognised that the assumption of many parallels cannot be convicted of inconsistency simply on the grounds of its generating strange results. Their work gave rise to the first non-Euclidean geometry. Later in the nineteenth century, Bernard Riemann (1826–66) saw how to modify the Euclidean postulates so as to allow for a consistent geometry based on the assumption of no parallels, and he provided a general approach to geometry in terms of the theory of manifolds. Riemann's abstract treatment did much to allay the suspicions of those who worried that contradictions might appear in the later development of non-Euclidean geometry, and such concerns were finally put to rest through the construction of relative consistency proofs by Eugenio Beltrami and Felix Klein. Throughout this process there was a subtle shift in the status of geometry. For almost two millennia after Euclid, geometry had seemed to be a science based on immediate and indubitable insights into the structure of space. After Gauss, Bolyai, Lobatschevsky and Riemann, Euclidean geometry appeared as one among many possible geometrical systems. The mathematical task seemed to be that of investigating the systems, while it would be left to the physicist to decide which of them correctly characterises space.

A similar assessment resulted from the late-eighteenth-century assimilation of the complex numbers. Jean Argand, Caspar Wessel and Gauss independently demonstrated that the algebra of complex numbers can be represented geometrically by identifying numbers with points on the two-dimensional plane. William Rowan Hamilton (1805–65) subsequently provided an algebraic interpretation in which complex numbers were taken to be ordered pairs of real. Seeking three-dimensional analogues of complex numbers, Hamilton formulated the algebra of quaternions, in which multiplication is non-commutative. His work in algebra, together with the nineteenth-century emergence of the concept of an abstract group from the theory of equations, fostered the belief in alternative algebraic structures, and suggested that mathematics consists in investigating these structures through axioms that uniquely characterise them.

The story of the foundations of analysis is more complex. Early in the nineteenth century, mathematicians working on problems generated by theoretical physics discovered that the lack of adequately-formulated central concepts of analysis handicapped their research. In his *Cours d'Analyse* of 1821, Augustin-Louis Cauchy (1789–1857) attempted to remedy the situation by providing definitions of the concepts of continuity, derivative and convergence in terms of the algebraic notion of limit. While Cauchy's treatment enabled him to give more comprehensible derivations of some of the main results of differential and integral calculus, it was not flawless. In particular, Cauchy endeavoured to resolve a controversy generated by the work of Joseph Fourier (1768–1830), who had claimed that arbitrary functions can be represented by trigonometric series. Opposing Fourier, Cauchy maintained that the sum of a convergent series of continuous functions is continuous (so that no convergent series of continuous functions can sum to a discontinuous function). As early as 1826, Niels Abel discovered that Cauchy's claim was false by finding a counter-example, but it took another three decades for mathematicians to explain in detail exactly how Cauchy had erred.

From this explanation mathematicians developed the theory of multiple limits and used it to distinguish between uniform and non-uniform continuity. Building on earlier work of Peter Lejeune Dirichlet and Philip Seidel, Karl Weierstrass (1815–97) provided the systematic treatment in the lectures that he gave in Berlin after his inauguration in 1856. Weierstrass's inaugural address made it apparent that he had been led to consider the foundations of analysis by a desire to pursue technical work in analysis (specifically to complete the theory of elliptic functions). This desire could only be satisfied after he had attained a clearer view of the central concepts of analysis; yet many years passed before he could devote himself to that project.

However, not all those who pursued the foundations of analysis did so as a prerequisite for solving a technical problem. Cauchy's contemporary, the Czech philosopher and mathematician Bernard Bolzano (1781–1848), explicitly sought the reform of analysis so as to uncover the proper explanation for the main results of the discipline. Although, like Cauchy, Bolzano failed to establish limit-existence theorems, his work prefigured subsequent studies of the structure of the continuum. Thirty years later, Richard Dedekind (1831–1916) was similarly moved by the need to make the principles of analysis comprehensible in attempting to define the continuity of the real line. He proposed that real numbers 'correspond to'

Dedekind sections of the rationals which made explicit, for the first time, the connection between the theorems of analysis and claims about infinite sets that Bolzano had already sensed. By the 1870s it had become obvious that Weierstrassian analysis was dependent on an explicit characterisation of the real numbers, and the available proposals – due to Dedekind, Charles Meray, Edward Heine and Georg Cantor (1845–1918) – all involved reference to infinite sets.

Some mathematicians suspected the concept of an arbitrary infinite set. Leopold Kronecker (1823–91) campaigned against the set-theoretic definitions of the real numbers used by his colleague Weierstrass, the set-theoretic constructions of Dedekind and, most of all, against Cantor's efforts to articulate the principles of transfinite set theory. Despite his objections, mathematicians faced with the choice between the strict arithmetisation of analysis favoured by Kronecker and the Weierstrassian reconstruction, recognised that arithmetisation could yield only fragments and they therefore opted for Weierstrass. Gradually, too, the power of Cantor's ideas became apparent. So the focus of foundational studies shifted from the justification of the principles of the calculus to the elaboration of set theory.

Meanwhile, significant changes occurred both in other parts of mathematics and in neighbouring areas of philosophy. After a placid history of over two hundred years, logic obtained new life in the nineteenth century, initially through the work of Bolzano and later through the researches of George Boole and Charles Sanders Peirce. But the major transformation was effected by Gottlob Frege (1848–1925). Originally trained as a mathematician, Frege came to believe that the search for rigour that had culminated in Weierstrassian analysis led to the investigation of the concepts of the arithmetic of the natural numbers. Influenced by Leibniz's claim that mathematics is a branch of logic, Frege set out to show that the concepts of arithmetic could be defined in terms of logical notions, and that the first principles of arithmetic derived from the laws of logic. To pursue his logicist programme, Frege needed a formal system of logic. Part of this was given in his *Begriffsschrift* (*Concept-script*) of 1879, and later extended in the two-volume *Grundgesetze der Arithmetik* (*Foundations of arithmetic,* 1893, 1903). In these works Frege developed a substantial part of modern formal logic.

Neither Frege's logical system nor his defence of logicism was positively received. Although *Begriffsschrift* was reviewed by several eminent logicians, few of Frege's contemporaries perceived that the daunting notation concealed a revolution in logic. Moreover, Frege had overestimated the kinship between the logicist programme and the enterprise of exposing the foundations of analysis. Mathematicians had chiefly been seeking problem-solving foundations; occasionally - as with Dedekind and Bolzano for explanation-yielding foundations. What Frege promised to deliver seemed less relevant: a more philosophical inquiry into the grounds of our certainty in elementary arithmetic. If there were residual foundational disputes, then the mathematicians took them to lie in the discussions about infinite sets; discussions to which Frege 's investigations did not seem pertinent.

However, some of Frege's contemporaries were also concerned to explore both logic and the foundations of elementary mathematics. In a little-appreciated monograph, *Was Sind und was Sollen die Zahlen*?, published in 1888, Dedekind also attempted a reconstruction of natural-number arithmetic and formulated the laws that characterise the natural numbers (the so-called Peano postulates). His motivations seem different from Frege's, in that Dedekind appears to have wanted explanation-yielding foundations and to have been indifferent to the philosophical tradition of searching for certainty-conferring foundations; a tradition linking Leibniz, Kant and Frege. Perhaps this explains why Dedekind was not concerned to specify the principles of his system with the precision and detail that Frege demanded, and why, in consequence, Frege failed to appreciate what he had accomplished.

Towards the close of the nineteenth century, Frege's efforts began to intertwine with those of other scholars. Building on earlier work by Moritz Pasch, who had perceived that Euclid's systematisation of geometry actually employs more premisses than are made explicit, David Hilbert (1862–1943) provided an axiomatisation of Euclidean geometry in which the deficiencies were repaired. He was also able to demonstrate that his system was consistent, relative to arithmetic. This success, together with the growing emphasis on mathematics as the study of axiomatic systems and the relative consistency proofs for non-Euclidean geometries, led Hilbert to formulate the problem of the foundations of mathematics in a way that accorded with the new image of mathematics. If mathematics is seen as the study of axiomatic systems, then the task of the foundationalist is to show that the systems are absolutely consistent.

That task became more urgent and Frege's logical investigations appeared more pertinent after Frege received a letter from Bertrand Russell (1872–1970) in 1902, which pointed out that his proposed system of logic was inconsistent. The problem affected not only Frege's logic but also Cantor's system of transfinite set theory, and it is most easily understood by reference to the latter. Cantor's system allowed that there should be a set of exactly those things that satisfy any arbitrarily chosen predicate (Frege's system contains an analogue of this principle). Russell pointed out that it follows that there is a set consisting of exactly those things that are not members of themselves. Call this set *R*. If *R* is not a member of itself then it is one of those things that are not members of themselves, and thus is a member of *R*. Conversely, if *R* is a member of itself then *R* belongs to the set whose members are just those things that are not members of themselves, so that *R* is not a member of itself after all. Since it must be the case that *R* belongs to itself or *R* does not belong to itself, we have a genuine contradiction.

Russell's paradox struck at the heart of Frege 's system because without a principle permitting the transition from arbitrary predicates to sets, (or more exactly, to what he called 'extensions of concepts') Frege saw no way to establish the existence of infinitely many objects and thereby to generate the usual theorems about the sequence of natural numbers. However, logicism was not the only victim. Mathematicians quickly recognised that the theorems of Weierstrassian analysis depended on the assumed existence of infinite sets, and the most natural way of guaranteeing existence was to adopt Cantor's principle. Evidently this had to be restricted in some way to save classical analysis. Russell's paradox – and other related paradoxes of set theory thus provoked several different types of foundational studies. Foundational work was required to solve pressing problems, to explain why classical mathematics seemed to work so well, and, for those steeped in the philosophical tradition, to confer on mathematics the certainty that had been called into question.

The search for indubitable foundations prompted three main philosophical programmes. Logicists struggled to find principles that could plausibly be taken for principles of logic, that would avoid the contradictions of Frege's system and that would suffice for the derivation of theorems of arithmetic and analysis. The most prominent achievement of logicism was the three volume *Principia Mathematica* (1910–13), jointly written by Russell and Alfred North Whitehead. Hilbert's profound essays on the foundations of mathematics outlined a different approach. Attuned to the new image of mathematics as the study of axiomatic systems, Hilbert suggested that certainty could again be assured by using an indubitable part of mathematics (roughly, primitive recursive arithmetic) to establish the consistency of the systems that are employed. With great ingenuity, Hilbert showed how formal systems could themselves be studied as mathematical objects, and how the statement of the consistency of these systems could be formulated as a mathematical assertion. His formalist programme was pursued with partial success into the 1930s.

A more radical foundational position was the intuitionism of Luitzen Brouwer (1881–1967). Taking up the complaints of Kronecker, Brouwer argued that the message of the paradoxes is that assertions of classical mathematics lack clear meaning. To talk of the existence of mathematical entities independent of the human power to construct such entities is to lapse into nonsense. The appropriate corrective is to require that any assertion of the existence of a mathematical entity should be based on a procedure for constructing it. This demand implies that the law of the excluded middle can no longer have unrestricted validity. Brouwer and his successors have laboured with considerable ingenuity to show that intuitionist analogues of classical theorems can be obtained. However, intuitionist analysis has proved considerably less powerful than its classical counterpart.

Few mathematicians and philosophers have been convinced by the complaint that the extra content of classical mathematics is rococo excess, and critics have objected that the restrictions imposed by intuitionists are arbitrary and unmotivated. Hilbert's formalist programme came to grief in a more dramatic way. In 1930, Kurt Godel (1906–78) proved a pair of theorems which effectively deny the possibility of finding the kind of consistency proof that Hilbert sought. In a slightly strengthened version, due to Barkley Rosser, Godel 's first incompleteness theorem states that, in any consistent formal system adequate for arithmetic, there will be a statement *A* such that neither *A* nor its negation is provable in the system. The second theorem asserts the unprovability of a statement expressing the consistency of the system (provided that the system is consistent). Although the consistency of formal systems adequate for elementary additive and multiplicative arithmetic can be proved by employing stronger methods than those that Hilbert allowed (as demonstrated by Jacques Herbrand and Gerhard Gentzen) the employment of these methods does not accord with the standards of indubitability and consistency that gave epistemological point to Hilbert's programme.

The fate of logicism was more complex. In their reconstruction of arithmetic and analysis, Russell and Whitehead were forced to introduce axioms whose credentials as logical principles were suspect. By the 1920s there was widespread agreement that the axioms of choice, infinity and replacement cannot be defended as laws of logic, and the champions of logicism began consequently to transform their position. They drew inspiration from some of the ideas of Hilbert, Rudolf Carnap (1891–1970) and other logical positivists who proposed in the 1930s that the axioms of a branch of mathematics are implicit definitions of the expressions occurring in them. In this way, mathematical knowledge starts with laws of logic and implicit definitions (which are indubitable in that our grasp of the meanings of the pertinent expressions suffices for our knowledge of the truth of the statements) and proceeds by deduction. This metamorphosis of logicism was severely criticised in a series of essays by Willard Van Orman Quine who advanced penetrating objections to the conception of truth-by-virtue-of-meaning on which the new logicism depended. By 1960, logicism, like the other two programmes that strove for certainty-conferring foundations, had been virtually abandoned.

But Russell's paradox did not only inspire the three philosophical enterprises whose history we have been tracing. The mathematical task of restricting the principle of Cantorian set theory that generated the troubles was undertaken by Ernst Zermelo (1871–1956), who formulated the iterative conception of set. Zermelo's set theory, extended by Abraham Fraenkel, sufficed not only for classical analysis but also for the description of those algebraic structures which have been vigorously studied by twentieth-century mathematicians. For the practising mathematician, the 'crisis' in the foundations of mathematics caused by the paradoxes of set theory has long been dissipated. The prevailing attitude, expressed eloquently by the group of mathematicians who write under the pseudonym Nicolas Bourbaki, is that the set theories of Zermelo and others provide a language in which to characterise and investigate the structures that are the proper object of mathematical research. This pragmatic attitude accords with that of Zermelo himself, who claimed explicitly that the credibility of his axioms depends not on their immediate indubitability, but on their ability to systematise and explain the results of classical mathematics.

From the point of view of mathematics, the 'foundational' subjects of logic and set theory have become parts of technical mathematics in their own right. In the early decades of this century, the formal logic of Frege and Russell was liberated from the assumption that logical statements have a fixed domain of discourse. Partly under the influence of the work of Peirce and Ernst Schroder, and partly through the investigations of Hilbert, the modern conception of an interpretation and a model was developed. Building on the work of Leopold Lowenheim, Thoralf Skolem established the important limitative theorem that any countable set of first-order sentences with an infinite model has a denumerable model. Godel's incompleteness theorems paved the way for related results by Alfred Tarski on the concept of truth, and by Alonzo Church on the undecidability of first-order logic. Godel's work was also seminal in the development of proof theory and recursive function theory. Since the Second World War, the theory of models has been further pursued and has established important connections with parts of algebra. One of its most celebrated accomplishments has been Abraham Robinson's development of non-standard analysis, and his revival of the long-derided notion of the infinitesimal. Finally, mathematicians have further articulated transfinite set theory; the investigations of Godel and Paul Cohen have established the independence of Cantor's continuum hypothesis from the axioms of set theory and the independence of the axiom of choice from the remaining axioms.

These accomplishments are probably best seen as the fruitful articulations of those parts of mathematics which at one time earned the title 'foundational'. While contemporary philosophers continue to seek certainty-conferring foundations, their investigations now seem quite divorced from the technical mathematical work that interests professional mathematicians. Some philosophers have taken inspiration from remarks by Godel, and have suggested that mathematics describes a universe of abstract objects sets and that the first principles describing this universe 'force themselves on us' as true. Others have offered ingenious attempts to revive logicism, to reformulate intuitionism, or to find a viable descendant of Hilbert's programme. But it is far from clear that any of these endeavours can answer to the original epistemological purpose of setting mathematics upon indubitable and immediate first principles.

Indeed, in the wake of Quine's influential critique of positivism, several philosophers, most notably Hilary Putnam and Imre Lakatos, have argued that the search for certainty-conferring foundations rests on a mistake. In Putnam's phrase, 'mathematics neither has nor needs foundations'. The present essay endorses this perspective, and while many philosophers have tried to read the history of the foundations of mathematics as exhibiting attempts by mathematicians to discover certainty-conferring foundations, my account portrays that history in terms of successes in obtaining problem-solving and explanation-yielding foundations. If mathematicians today evince any concern for foundations, they will concentrate on explanations and solutions to problems. Where there is no need for explanation and no urgent problem that requires conceptual clarification they will view philosophical inquiries as irrelevant and pointless. According to my version of the history of the foundations of mathematics, their attitude should come as no surprise. *Plus ça change, plus c'est la même chose.*

Bibliography

Babbage, Charles, *A Comparative View of the Various Institutions for the Assurance of Lives* (1826; reprinted New York: Augustus M. Kelley, 1967).

Babbage, Charles, *On the Economy of Machinery and Manufactures*, 3d ed. (London: Charles Knight, 1833).

Bailey, Arthur, and Archibald Day, "On the Rate of Mortality amongst the Families of the Peerage," *Assurance Magazine*, 9 (1860–61), 305–326.

Cairns, John Jr., and James R. Pratt, "The Scientific Basis of Bioassays," *Hydrobiologia*, 188/189 (1989), 5–20.

Calhoun, Daniel, *The American Civil Engineer: Origins and Conflict* (Cambridge, Mass.: MIT Press, 1960).

Calhoun, Daniel, *The Intelligence of a People* (Princeton, N.J.: Princeton University Press, 1973).

Campbell-Kelly, Martin, "Large-Scale Data Processing in the Prudential, 1850- 1930," *Accounting, Business, and Financial History*, 2 (1992), 117–139.

Daston, Lorraine, "Objectivity and the Escape from Perspective," *Social Studies of Science*, 22 (1992), 597–618.

Daston, Lorraine, "The Ideal and Reality of the Republic of Letters in the Enlightenment," *Science in Context*, 4 (1991), 367–386.

Daston, Lorraine, *Classical Probability in the Enlightenment* (Princeton, N.J.: Princeton University Press, 1988).

Eckstein, Otto, *Water-Resource Development: The Economics of Project Evaluation* (Cambridge, Mass.: Harvard University Press, 1958).

Ekelund, Robert B., and Robert F. Hébert, "French Engineers, Welfare Economics, and Public Finance in the Nineteenth Century," *History of Political Economy*, 10 (1978), 636–668.

Elwitt, Sanford, *The Making of the Third Republic: Class and Politics in France, 1868–1884* (Baton Rouge: Louisiana State University Press, 1975).

Elwitt, Sanford, *The Third Republic Defended: Bourgeois Reform in France, 1880–1914* (Baton Rouge: Louisiana State University Press, 1986).

Finkelstein, Michael, and Hans Levenbach, "Regression Estimates of Damages in Price-Fixing Cases," in DeGroot et al., *Statistics*, 79–106.

Finney, Donald J., *Statistical Method in Biological Assay*, (London: Charles Griffin and Co., 1952).

Fischhoff, Baruch, et al., *Acceptable Risk* (Cambridge, U.K.: Cambridge University Press, 1981).

Fish, Stanley, *Is There a Text in This Class? The Authority of Interpretive Communities* (Cambridge, Mass.: Harvard University Press, 1980).

Garnier, Joseph, "Sur les frais d'entretien des routes en empierrement," *Annales des Ponts et Chaussées* [2], 10 (1845), 146–196.

Geiger, Reed, "Planning the French Canals: The 'Becquey Plan' of 1820- 1822," *Journal of Economic History*, 44 (1984), 329–339.

Geison, Gerald L., "'Divided We Stand': Physiologists and Clinicians in the American Context," in Morris J. Vogel and Charles Rosenberg, eds., *The Therapeutic Revolution: Essays in the Social History of American Medicine* (Philadelphia: University of Pennsylvania, Press, 1979), 67–90.

Geison, Gerald L., "Scientific Change, Emerging Specialties, and Research Schools," *History of Science*, 19 (1981), 20–40.

Hacking, Ian, "Statistical Language, Statistical Truth, and Statistical Reason: The SelfAuthentification of a Style of Scientific Reasoning," in Ernan McMullin, ed., *The Social Dimensions of Science* (Notre Dame: University of Notre Dame Press, 1992), 130–157.

Hacking, Ian, "Telepathy: Origins of Randomization in Experimental Design," *Isis,* 79 (1988), 427–451.

Hacking, Ian, *The Taming of Chance* (Cambridge, U.K.: Cambridge University Press, 1990).

Ijiri, Yuji, and Robert K. Jaedicke, "Reliability and Objectivity of Accounting Measurements," *Accounting Review,* 41 (1986), 474–483.

Ingrao, Bruna, and Giorgio Israel, *The Invisible Hand: Equilibrium in the History of Science,* Ian McGilvray, trans. (Cambridge, Mass.: MIT Press, 1987).

Jacobs, Stuart, "Scientific Community: Formulations and Critique of a Sociological Motif," *British Journal of Sociology,* 38 (1987), 266–276.

Jaffé, William, ed., *Correspondence of Léon Walras and Related Papers,* 3 vols. (Amsterdam: North Holland, 1965).

Jasanoff, Sheila, "The Misrule of Law at OSHA," in Dorothy Nelkin, ed., *The Language of Risk* (Beverly Hills: Sage, 1985), 155–178.

Keyfitz, Nathan, "The Social and Political Context of Population Forecasting," in Alonso and Starr, *Politics of Numbers,* 235–258.

Kindleberger, Charles P., "Technical Education and the French Entrepreneur," in E. C. Carter et al., eds., *Enterprise and Entrepreneurs in Nineteenth- and Twentieth-Century France* (Baltimore: Johns Hopkins University Press, 1976).

La Gournerie, Jules de, "Essai sur le principe des tarifs dans l'exploitation des chemins de fer" (1879), in La Gournerie, *Etudes économiques.*

La Gournerie, Jules de, *Etudes économiques sur l'exploitation des chemins de fer* (Paris: GauthierVillars, 1880).

Megill, Allan, ed., *Rethinking Objectivity,* special issue of *Annals of Scholarship,* 8 (1991), parts 3–4, and 9 (1992), parts 1–2.

Mehrtens, Herbert, *Moderne – Sprache – Mathematik: Eine Geschichte des Streits um die Grundlagen der Disziplin und des Subjects formaler Systeme* (Frankfurt am Main: Suhrkamp Verlag, 1990).

Mellet and Henry, MM., *L'arbitraire administratif des ponts et chaussées dévoilé aux chambres* (Paris: Giraudet et Jouaust, 1835).

Ménard, Claude, *La formation d'une rationalité économique: A. A. Cournot* (Paris: Flammarion, 1978).

Nelson, John S., Allan Megill, and Donald McCloskey, eds., *The Rhetoric of the Human Sciences* (Madison: University of Wisconsin Press, 1987).

Owens, Larry, "Patents, the 'Frontiers' of American Invention, and the Monopoly Committee of 1939: Anatomy of a Discourse," *Technology and Culture,* 32 (1992), 1076–1093.

Palmer, Robert R., *The Age of the Democratic Revolution,* 2 vols. (Princeton, N.J.: Princeton University Press, 1959–1964).

Parker, R. H., *The Development of the Accountancy Profession in Britain in the Early Twentieth Century* (London: Academy of Accounting Historians, 1986).

Poincare H. (1898). *Celestial mechanics* (Vol. III, p. 1059, sec. 397).

Smith, Roger, and Brian Wynne, "Introduction," to Smith and Wynne, *Expert Evidence,* 1–22.

Smith, Roger, eds., *Expert Evidence: Interpreting Science in the Law* (London: Routledge, 1989).

Temin, Peter, ed., *Inside the Business Enterprise* (Chicago: University of Chicago Press, 1991).

Terrall, Mary, "Maupertuis and Eighteenth-Century Scientific Culture" (Ph.D. dissertation, University of California, Los Angeles, 1987).

Tsuda I. (1991). "Chaotic itinerancy as a dynamical basis of hermeneutics in brain and mind". *World Futures, 32,* 167-184.

Weisbrod, Burton A., "Costs and Benefits of Medical Research: A Case Study of Poliomyelitis," in *Benefit-Cost Analysis: An Aldine Annual, 1971* (Chicago: Aldine-Atherton, 1972), 142–160.

Weiss, John H., *The Making of Technological Man: The Social Origins of French Engineering Education* (Cambridge, Mass.: MIT Press, 1982).

Zak M. (1991). "Terminal chaos for information processing in neurodynamics". *Biological Cybernetics, 64,* 343-351.

Index